T0093491

Cyber Physical Systems

Cyber physical system (CPS) is an integration of computation, networking, and physical processes: the combination of several systems of different natures whose main purpose is to control a physical process and, through feedback, adapt itself to new conditions in real time. *Cyber Physical System: Concepts and Applications* includes in-depth coverage of the latest models and theories that unify perspectives.

It expresses the interacting dynamics of the computational and physical components of a system in a dynamic environment.

Key Features

- Covers automatic application of software countermeasures against physical attacks and impact of cyber physical system on Industry 4.0

- Explains how formal models provide mathematical abstractions to manage the complexity of a system design

- Offers a rigorous and comprehensive introduction to the principles of design, specification, modeling, and analysis of cyber physical systems

- Discusses the multiple domains where cyber physical system has a vital impact and provides knowledge about the use of specialized models that serve as mathematical abstractions to control the complexities of a system's architecture

- Provides the rapidly expanding field of cyber physical systems with a long-needed foundational text by a panel of domain experts

This book is primarily aimed at advanced undergraduates and graduates of computer science. Engineers will also find this book useful.

Chapman & Hall/CRC Cyber-Physical Systems

SERIES EDITORS:

Jyotir Moy Chatterjee
Lord Buddha Education Foundation, Kathmandu, Nepal

Vishal Jain
Sharda University, Greater Noida, India

Cyber-Physical Systems: A Comprehensive Guide
By: Nonita Sharma, L K Awasthi, Monika Mangla, K P Sharma, Rohit Kumar

Introduction to the Cyber Ranges
By: Bishwajeet Pandey and Shabeer Ahmad

Security Analytics: A Data Centric Approach to Information Security
By: Mehak Khurana & Shilpa Mahajan

Security and Resilience of Cyber Physical Systems
By: Krishan Kumar, Sunny Behal, Abhinav Bhandari and Sajal Bhatia

Cyber Security Applications for Industry 4.0
By: R Sujatha, G Prakash, Noor Zaman Jhanjhi

Cyber Physical Systems: Concepts and Applications
By: Anupam Baliyan, Kuldeep Singh Kaswan, Naresh Kumar, Kamal Upreti and Ramani Kannan

For more information on this series please visit: https://www.routledge.com/Chapman--HallCRC-Cyber-Physical -Systems/book-series/CHCPS?pd=published,forthcoming&pg=1&pp=12&so=pub&view=list?pd=published,forth- coming&pg=1&pp=12&so=pub&view=list

Cyber Physical Systems

Concepts and Applications

Edited by
Anupam Baliyan, Kuldeep Singh Kaswan,
Naresh Kumar, Kamal Upreti, and Ramani Kannan

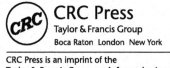

CRC Press
Taylor & Francis Group
Boca Raton London New York

CRC Press is an imprint of the
Taylor & Francis Group, an **informa** business

A CHAPMAN & HALL BOOK

First edition published 2023
by CRC Press
6000 Broken Sound Parkway NW, Suite 300, Boca Raton, FL 33487-2742

and by CRC Press
4 Park Square, Milton Park, Abingdon, Oxon, OX14 4RN

CRC Press is an imprint of Taylor & Francis Group, LLC

Library of Congress Cataloging-in-Publication Data

Names: Baliyan, Anupam, 1976- editor. | Kaswan, Kuldeep Singh, editor. | Kumar, Naresh, 1983- editor. | Upreti, Kamal, editor. | Kannan, Ramani, editor.
Title: Cyber physical systems: concepts and applications / edited by Anupam Baliyan, Kuldeep Singh Kaswan, Naresh Kumar, Kamal Upreti, Ramani Kannan. Other titles: Cyber physical systems (CRC Press: 2023)
Description: First edition. | Boca Raton : Chapman & Hall/CRC Press, 2023. | Series: Chapman & Hall/CRC cyber-physical systems | Includes bibliographical references and index. |
Identifiers: LCCN 2022041539 (print) | LCCN 2022041540 (ebook) | ISBN 9781032116044 (hardback) | ISBN 9781032420653 (paperback) | ISBN 9781003220664 (ebook)
Subjects: LCSH: Cooperating objects (Computer systems).
Classification: LCC TJ213 .C8845 2023 (print) | LCC TJ213 (ebook) | DDC 006.2/2--c23/eng/20221116
LC record available at https://lccn.loc.gov/2022041539
LC ebook record available at https://lccn.loc.gov/2022041540

ISBN: 978-1-032-11604-4 (hbk)
ISBN: 978-1-032-42065-3 (pbk)
ISBN: 978-1-003-22066-4 (ebk)

DOI: 10.1201/9781003220664

Typeset in Palatino
by Deanta Global Publishing Services, Chennai, India

Contents

Preface

Cyber physical system (CPS) is an integration of computation, networking, and physical processes: the combination of several systems of different natures whose main purpose is to control a physical process and, through feedback, adapt itself to new conditions in real time. This book includes an in-depth coverage of the latest models and theories that unify perspectives, expressing the interacting dynamics of the computational and physical components of a system in a dynamic environment.

This book focuses on the new design, analysis, and verification tools that embody the scientific principles of CPS and incorporate measurement, dynamics, and control. It also covers the application of CPS in numerous sectors, including agriculture, energy, transportation, building design and automation, healthcare, and manufacturing. This book presents an in-depth review of the state of the art of cyber physical systems and their applications, reviews the design challenges of CPS, and evaluates their impact on systems and software engineering.

The book addresses foundational issues central across CPS applications, including system design: how to design CPS to be safe, secure, and resilient in rapidly evolving environments; system verification: how to develop effective metrics and methods to verify and certify large and complex CPSs; real-time control and adaptation: how to achieve real-time dynamic control and behavior adaptation in diverse environments, and role of CPS in Industry 4.0.

About the Editors

Anupam Baliyan serves as an Additional Director (Engineering) in the Department of Computer Science and Engineering, University Institute of Engineering, Chandigarh University, Mohali, Punjab. He has more than 22 years of experience in academics. Baliyan obtained his MCA from Gurukul Kangari University, Haridwar, and his MTech (CSE) and PhD (CSE) from Banasthali University, Rajasthan. He has published more than 30 research papers in various international journals indexed at Scopus and ESI. He is a lifetime member of CSI and ISTE. He has chaired many sessions at international conferences across India. Baliyan has published some edited books and chapters. He is also the assistant editor of some journals that are Scopus indexed. His research area includes algorithms, machine learning, wireless networks, and AI.

Kuldeep Singh Kaswan is a Professor in the School of Computing Sciences and Engineering, Galgotias University, Uttar Pradesh. He received a doctorate in Computer Science under the Faculty of Computer Science at Banasthali Vidyapith, Rajasthan. He obtained his master's in Computer Science and Engineering from Choudhary Devi Lal University, Sirsa (Haryana). He is also a member of the Computer Science Teacher Association (CSTA), New York, USA; International Association of Engineers (IAENG), Hong Kong; International Association of Computer Science and Information Technology (IACSIT), USA; professional member of Association of Computing Machinery, USA, and IEEE; and life member of Computer Society of India, India. His areas of interest include IoT, machine learning, and soft computing. He has a number of publications in international/national journals and conferences to his credit.

Naresh Kumar is a Professor in the Department of Computer Science and Engineering, GL Bajaj Institute of Technology and Management, Greater Noida, Uttar Pradesh, India. He has 15 years of experience in teaching and research. He received his PhD in Computer Engineering from the University Institute of Engineering and Technology, Maharishi Dayanand University, Rohtak, Haryana. He completed his MTech in Computer Science and Engineering from Kurukshetra University, Kurukshetra, Thanesar, Haryana, in 2004. He has presented several papers at international conferences and published work in peer-reviewed and Science Citation Index (SCI) journals. Kumar has also authored a book titled *Objective Computer Science: Training Minds and Creating Opportunities* (Inkart Publishing). He is an active reviewer for many top journals of IEEE, IET, and Springer. He is also a Cisco-certified trainer for Cisco Network Security, CCNA, and CyberOps. His area of research covers information extraction, big data analytics, and computer networking.

Kamal Upreti is an Associate Professor in the Department of Computer Science & Engineering, Dr. Akhilesh Das Gupta Institute of Technology & Management (formerly NIEC), affiliated with Guru Govind Singh Indraprastha University, Delhi. He is a corporate trainer in HCL in the field of cyber security and data science. He completed his BTech (hons) from UPTU, MTech (gold medalist) from Galgotias University, PGDM (Executive) from IMT Ghaziabad, and PhD in Computer Science and Engineering from OPJ University.

He has obtained 30+ patents and published 30+ books, 32+ magazine issues, and 40+ research papers in various international conferences and reputed journals. His areas of interest are physics, cyber security, machine learning, wireless networking, embedded system, and cloud computing. He has years of experience in corporate and teaching experience in engineering colleges.

He worked with HCL, NECHCL, *The Hindustan Times*, Dehradun Institute of Technology, and Delhi Institute of Advanced Studies, with more than 15+ years of enriching experience in research, academics, and corporate sector. He also worked with NECHCL in Japan in the project "Hydrastore," funded by a joint collaboration between HCL and NECHCL. Upreti worked on a government project—"Integrated Power Development Scheme (IPDS)"—launched by the Ministry of Power, Government of India, with the objective of strengthening the sub-transmission and distribution network in the urban areas. He has served as a session chairperson at national, international conferences and as a keynote speaker in various platforms such as skill-based training, corporate trainer, guest faculty, and faculty development programs. He has been awarded with the titles of the best teacher, best researcher, extra-academic performer, and was also a gold medalist in his MTech program.

He has published patents, books, magazine issues, and research papers in various national, international conferences and peer-reviewed journals. His research area includes cyber security, data analytics, wireless networking, embedded system, neural network, and artificial Intelligence.

Ramani Kannan obtained his PhD in power electronics and drives, Anna University, Chennai; ME in power electronics and drives, Anna University; and BE in electronics and communication, Bharathiyar University, Coimbatore.

Kannan has teaching, administration, and research experience in the past decade working in K.S. Rangasamy College of Technology, Tiruchengode, Tamil Nadu. He is currently working in Department of Electrical and Electronics Engineering at Universiti Teknologi PETRONAS, Perak, Malaysia.

Career Achievements:

- Career Award for Young Teachers (CAYT) from AICTE, New Delhi in the year 2013–2014
- Best Teacher Award (2010–2011) at K.S. Rangasamy College of Technology, Tiruchengode
- Outstanding Teacher Award for 100% results produced: 13 times in university and end semester results
- Outside Interaction Editor-in-Chief, *Journal of Asian Scientific Research* (Online ISSN: 2223-1331 Print ISSN: 2226-5724); REGIONAL EDITORS SOUTH-ASIA *International Journal of Computer Aided Engineering and Technology*, Inderscience publisher (Scopus index) ISSN online: 1757-2665

List of Contributors

Abhishek Sharma
Vikram University, India

Anupam Baliyan
Chandigarh University, Punjab

Amit Pandey
Bule Hora University, Ethiopia

Avadhesh Kumar
Galgotias University, India

Bharti Nagpal
Netaji Subhas University of Technology,
India

Chhavi Sharma
Abdul Kalam Technical University,
India

Himanshu
Netaji Subhas University of Technology,
India

Javed Miya
Galgotias College of Engineering and
Technology, India

Jagjit Singh Dhatterwal
Koneru Lakshmaiah Education
Foundation, India

Kamal Upreti
Dr. Akhilesh Das Gupta Institute of
Technology and Management, India

Keerti Dixit
Vikram University, India

Kuldeep Kaswan
Galgotias University, India

Naresh Kumar
Galgotias University, India

Nitin
PDM University, India

Pawan Kumar singh
Galgotias University, India

Pragya Singh Tomar
Vikram University, India

Piyush Parkash
PDM University, India

Ram Shringar Rao
Netaji Subhas University of Technology,
India

Rajbala
Galgotias University, India

Rajesh Kumar
JECRC University, India

Sanjay Kumar
Galgotias University, India

Shobha Bhatt
Netaji Subhas University of Technology,
India

Shalini Singh
PDM University, India

Suyash Kumar Singh
University of Connecticut, USA

Sahil Kansal
Jagan Institute of Management Studies,
India

Umesh Kumar Singh
Vikram University, India

Vishnu Sharma
Galgotias University, India

1

Digital Flow-Based Cyber-Physical Microfluidic Biochips

Jagjit Singh Dhatterwal and Anupam Baliyan

CONTENTS

1.1 Introduction

A cyber-physical microfluidic biochip (CPMB) is essentially a fluid handling system at its heart. Microfluidic refers to the existence of microliters or smaller fluid sizes. Microfluidics apply more generally to various methods used to manipulate such small-scale fluids. Such developments are multidisciplinary and encompass, for one thing, the areas of polymer science, mechanics, electro-technology, biotechnology, and computer design. Microfluidic fluid processing is also called a "biochip" or a "microfluidic framework." "Microfluidic device" refers to electronic components, while "bio" refers to biochemical applications. Cyber physical system occurs when detectors, transducers, and controls are included to expand microfluidic biochip technologies significantly. All of this may together be

DOI: 10.1201/9781003220664-1

referred to as a microfluidic cyber-physical device or network, which, where applicable, is the language used. We can use CPMB to refer to the whole system, but in particular, one must be well prepared for the optimal device with all these modules: a lightweight and user-friendly platform for experimental procedures. For example, terms like micro-total-analytical systems (µTAS), lab-on-an-chip (LOC), and sampling-to-answer methods are often interchangeable terms with microfluidics.

In this part, we study the CPMB technologies. In this work, we are focusing on two strategies suggested that supply cyber-physical integration and design automation technologies, while there are several ways to control low amounts of fluid: digital microfluidic biochips (DMFBs) and fluidic microfluidic biochips (FMFBs).

1.2 What Is Microfluidics

Just the size of the controlled fluids is the literal description of microfluidics. The word in terms of application specifics is also vague since various technologies can conduct fluid manipulation on a small scale. The limitless advantages of the scale guided the invention of microfluidics into the sub-microliters field by conventional clinical laboratories:

- Sample and reagent use decreased: Several systems test samples or manipulate them with many reagents. The reduction in fluid intake to lower volumes inevitably reduces. This is appealing because ingredients can be hard to acquire and even more challenging to substitute samples.

- Improved output: A high-volume surface-to-area ratio allows greater temperature control, which results in increased resolution and sensitivity performance [1].

- Low power consumption: Low mass ensures that the movement of fluids needs less force. Tools configured for compact and remote operation or operation with low power consumption gain also enables a closer integration of the physical device, as it means less cell heating and less energy transfer. Surplus heating can also conflict with the testing process.

- Transportation: Another small-scale natural by-product. Unfortunately, the portability bottleneck has proven to be the supporting hardware for the microfluidic biochip operation [2].

- Physical impacts: Fluids in small sizes are subject to laminar flow instead of turbulent flow. This is to the benefit of the microfluidic biochip builder since fluid movements can be predicted and monitored more effectively. Surface/interface stress, capillary pressures, and fluid movement across channels dominate in the microscale [3].

- Automation: Typical lab practices require comprehensive human workflow interference, including installation and movement of measurements and reagents between different instruments. Scaling enables the convergence of several other device elements into one substratum, reducing the need for human interaction while still allowing increasingly sophisticated protocols to be implemented.

The important point [4] is about how molecular research, biodefense, molecular genetics, and microscopy applied to microfluidics as we know it today, from the fascinating topic of view of the growth of microfluidics. The fluid processing methods are substantially different, such as document [5] droplets and auditory activation, reaction kinetics [7]. These developments do not feature in this paper. Despite all this diversity, the subjects of the rest of this segment can still be discussed in the type of information system.

1.2.1 Significance of Microfluidic Biochips

The ideal version shape of a microfluidic system and its application in the real world are significantly different at the moment. The ideal can be thought of as a *Star Trek* diagnostics tricorder: portable, inexpensive, non-invasive, all of which can detect a specimen to provide the testing at quicker response times. The reality is that they are mainly cumbersome experimental interfaces. There are also certain implementations as a wick that require the proper functioning of several external laboratory instruments. This limits their use as natural laboratory on-chip platforms. Devices without this restriction are autonomous and can be labeled by their actuation mechanism [8].

- Passive, personality microfluidic biochips use capillary flow and colorimetric sensing systems with no international support capability. Examples of passive microfluidics and home blood transfusion are their versatility and precision have brought enormous, colossal success.
- Hand powered devices require human intervention, either pressed, piped, or squeezed blister packs, to supply the primary driver. Active systems use mechanical sensors, actuators, pumps, and controls to simplify fluid handling.

For example, many compute clusters use humans to pipette them into a cartridge and load them into an automatic processing system. Computer operating systems may have hybrid properties. The main subject of this book is microfluidic biochips that are involved. In some cases, we study them as they already exist now, but in others, we intend them to be completely self-contained structures into eventual deployment.

1.2.2 Intelligent Control Cyber-Physical Integration

A biochip includes only the substratum for manipulating fluids. This is sufficient to achieve the necessary flexibility for specific devices, such as those dependent on blood capillary acceleration. But cyber-physical integration, including the use of physiological, behavioral manipulation sensors, actuation, and communication systems, is essential for more group projects on DMFBs and FMFBs [9].

The sensor's feedback and monitoring will work together to execute scattering decisions in the provisional protocol, an equipment issue or a mistake like an inadequate mix [10], or carry out corrective measures. Cyber-physical convergence also allows for the communication of CPMBs to other computers. Transfer information recorded and edited on-field for review. This opens up known vulnerabilities and is the subject of the following chapter.

For example, sensors for CPMBs include spectroscopy detectors, cameras, and, at times, patented designs like Nanopore Technology from Oxford that detects shifts in the current as particles go through a nano-free hole. Sensors may be built into the biochip in specific ways. However, these elements and the interface require equipment greater than the size of the chip to be incorporated for the charge of fluids in the biosensor. This is due to the design of the commonly used techniques used in biochips. In contrast, initial experiments for the manufacture of silicon-borrowed microfluidics are now widely used in PDMS and low-cost PCB production of polydimethylsiloxane.

1.2.3 Techniques of CAD

- Growth of computer-aided control strategies, progressively advanced methods, and protocols for fluid handling have partly influenced the systems engineering community with tremendous achievements [11]. Manual design and monotonous operating solutions support some of these technologies, while others can be designed to enable high-performance, increased applications. The CAD community for microfluidics is too large to explain here and perhaps outside the reach of this text, but we make some significant contributions. Architectural synthesis: The challenge of creating a biochip architecture that meets those performance targets can be addressed with metaheuristic methods such as taboo searching and computational annealing [12].

- Fault-tolerance and reliability: A variety of failure modes affect microfluidic systems. Any of them may be promptly induced during daily use. Improved yield and stock life can be accomplished by improving stability in the design process [13, 14].

- Test: While complexities are permanently reduced, the difficulty for practical research in microfluidics remains unique to the VLSI test. The time variables are slower, and the trial phase may degrade consistency. The test methodologies comprised microfluidic biochip modeling as an analog sequence of signal processing and the usage of VLSI techniques with some minor alterations [15–18].

- Optimization of the protocol: While we always assume that the biochemist has a mentally prepared protocol for platform execution, this mechanism will be streamlined and optimized in many cases. For example, the before analysis procedure is the diluting and mixing sample fluid in many known quantities. Excess fluid intake and biochip deterioration are likely to occur [19–22].

1.2.4 Level of Design

- Currently, CPMBs are manufactured in vertical directions; the devices are designed and produced in the same workshop or plant. It is fair to anticipate that a longitudinal supply chain would be introduced with new technology comparable to the semiconductor industry. The modernization of biochip architecture is also possible when the parties no longer fight to build their fabrication equipment to design new methodological procedures.

- A microfluidic biochip can use one of the two concept flows described in Figure.1.1. In general-purpose movement, the bio code first describes the logical characterization of what the test is supposed to do. A high standard specifier using terminology such as BioCoder [23] is produced by the bio code, which includes a comprehensive definition of the test with standards download, remix, temperature, and identify. The top-level setup is subsequently sent to the biosensor manufacturers to aggregate the features and capabilities required for biochip production. The manufacturer uses a CAD tool provider's biochip supplier's equipment/software

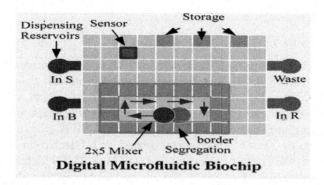

FIGURE 1.1
Design level of microfluidic biochips.

equipment. The hardware supplier offers production details such as the specific rules for gravure of electrodes, tubes, valves, pumps, and fluids, which can be comfortably regulated on the device. The hardware vendor serves the same purpose as the foundry in the IC design process and will provide the biochip designer with a product development package.

The vendor of CAD instruments writes a program for synthesis which transforms the test data into indications on biochip hardware. The concept is produced, installed, and submitted to the verification tester and the end customer. After the evaluation is completed, the unit must be collected in the plant that takes biological waste for handling and recovery since its useful life has been surpassed. An implementation biochip design flow is identical, but before being supplied to the manufacturing plant, design details should be sequenced by the biochip manufacturer.

1.2.5 Applications of Microfluidics

- Scientific research: This covers drug development and a range of "omics," including genomics and proteomics, some of the primary uses of microfluidics. Historically, these operations have only been confined to laboratories with many infrastructure and service expenditures. The use of microfluidics can accomplish high-performance sequencing in laboratories with small resources.

- Defense and community security: The detection of harmful chemicals and pathogens can be done using microfluidic instruments. This was one of the principal reasons microfluidic technologies were developed [4]. For example, surveillance of contaminants in the environment [24, 25], identification of bacteria or food protection antibiotics [26], and detection of biological and chemical weapons [27] are both microfluidic instruments designed for security applications. This unique feature is intriguing since it delivers a physical wellbeing measurement but is prone to dangerous usage.

Medical diagnosis: The capacity to quickly diagnose disorders and multichannel assays can boost holistic care and better quality of treatment, located in poor settings from outside medical clinics. In the sense of protection and confidence, we will review several of these implementations in the following paragraphs.

1.3 Manipulation of DMFB Encapsulation

The microfluidics class comprises electronic microfluidic biosensors that use isolated droplets to handle fluids [28, 29]. It is different from microfluidic droplets [6], which are instruments encapsulating target molecules in goblets. The distinction is that droplets are shuttled through reservoirs and channels to vibration microfluidics, and digital microfluidic biochips travel at distinct distances. The expression "DMFB" does not presume a single type of medical triggering but has become correlated with "electro-on-dielectric" (EWOD) over the last decades [30]. The configuration of a standard DMFB includes a designed electrode grid that is hydrophilic (Figure.1.2). Vessels and pipelines may also be modeled, and the square or rectangular form of the electrode's grids may be formed. If there is a top surface, it is mainly made of a translucent conduction material to observe drip motions and the electrode plate. When a voltage is applied to the electrodes, the EWOD is induced to goblets and may be used by fluids. Using a large-scale avoids evaporative fluid cooling; requiring droplets to move in a charged particle-like silicon oil may decrease the necessary switching frequency.

1.3.1 Electro Wetting-On-Dielectric-Based Microfluidic Chips

The Young–Lippmann equation can be used to model electro-on-dielectric [28]:

$$0rV2$$

$$\cos\theta = \cos\theta0 + \underline{} \tag{1.1}$$

$$2\gamma d$$

While θ is the oscillating angle and is the static contact angle of μL and the substratum, ±0 is the static pressure difference without stress, r is the dynamic dielectric viscosity, V is the voltage, μ is the filling media surface tension, and d is the dielectric thickness. The touch angle can be modified, and activities such as the supplying, transportation, mixing, and sweating of droplets can be performed by changing the voltage level and controlling the output of the DMFB. Subject to complicated biosensors, including in-vitro and random mutagenesis biosynthesis, these operations may be applied [28, 31]. In reality, industrial device designers describe the fundamental mechanics of EWOD droplet motion without affecting efficiency or functionality in any aspect directly. In general, a grid of electrodes is sufficient for a controller to transmit rising or falling voltage levels. The application guides particle migration to a dropstick electrode of lower power and the adjacent electrodes of high voltage. Application components are known as switching devices used to

FIGURE 1.2
Encapsulation of digital microfluidic biochips.

perform bioassay. The electronic circuits which are logically high (1) and conceptually low (0) can be delivered from the digital microcontroller to the DMFB through the right level charging facilities (as DMFBs usually need hundreds of voltage regulators almost always!). Applications of the DMFB are constrained in terms of the mechanics of the acceleration of the droplet and are in the order of hours to thousands of hertz in standard systems [28].

1.3.2 Generation of High Complexity in Microfluidics

- For even reasonable simplicity tests on small DMFB arrays, the difficulty of producing actuating sequences is unaffordable. To cope with this strain, the idea of high-level proliferation was introduced early from VLSI [11]. Generally, high-level DMFB flow begins with a bioassay requirement to be performed on chip. The figurative language of this requirement is highly detailed and is translated to the process synthesis applications. In the general case, three problems with NP-hard specialization are resolved.

- The planning shall determine the scientific method to meet the requirements of precedent resources [11]. Installations are the distribution for activity, such as combining physical biochip resources [13]. There are two different methods: free positioning, using FPGA floors scheduling algorithms [32], and virtual topology positioning [33, 34], which presupposes those operations can include only those chip areas. In terms of compact size and profitability, virtual topography also contributes to better quality performance.

- Droplet transfer routes from their dispatch ports to running modules must be determined and then to a waste port or an output terminal. VLSI scheduling algorithms, such as a Soukup labyrinth router [35], may be implemented, and multi-droplet routing can be compressed to simultaneous occurrence [36].

- The goal of automation is usually a time of fulfillment, efficiency, and testing [11, 37]. The same resources for compilation and routing within a DMFB are common for both of those steps to their VLSI equivalents. In VLSI fabrication, semiconductors and switches are employed to perform logic, whereas lines are used for interconnection. DMFBs utilize conductors both for the processing and transmission of liquids. It has been suggested to address alternative synthesis flows continuously in all transitions, often with higher effects of performance optimization [17, 38, 39].

1.3.3 Error Recovery in Microfluidic Chips

DMFBs are prone to a wide range of faults due to possible defects such as constant load and several mechanisms during bioassay [40]. Biosynthetic and checkpoint-dependent error recovery Bayesian inference was demonstrated as one of the most successful ways to address DMFB engineering's core shortcoming.

International border failure retrieval tracks the application of DMFBs in space and time at specific locations. If a mistake is discovered, it can redirect time to restore more efficiently than restart the test. CCD cameras are also used to provide a biochip controller with input from the sensor in real time. CCD cameras can be configured, and the biochip's status can be determined in random places and times. Controller-run software uses pattern-matching algorithms to collect images from the camera and derives droplet presentation, length, and concentration from implementing programs. The proposed models involve creating a map for the whole biochip and evaluating the position of enzymes with the most significant correlation. The association of a single droplet after the picture has been cropped to a certain point. While CCD scanning is versatile and reliable, it may not be used to test luminous reagents. As has recently been seen with reliability-hardening, we also notice that different findings suggest they can execute safety control signals.

1.3.4 Direct Addressing in DMFBs

The sample size used to drive a DMFB is one of the most significant contributors to costs and complexity. The earliest generic DMFBs needed one IO pin from the driver circuit, a system called "specific addresses." And for recent addition, this can quickly become unrealistic. For example, a proved biosensor DMFB for a moment in time testing demands over one thousand pins—simply adding to typical MCU packets the number of connections—for the use of direct management. Pin-constrained DMFBs lower the number of pins and limit the mobility of the drop of water. The final stage in the good innovative flow can be produced with pin-constrained DMFBs, or the overall biochip architecture can be allowed. The same immunological test is employed when the biochip's 64 less pins are used to transmit more than 1,000 terminals.

1.3.5 Commercialization of Digital Microfluidic Technology

- The authors know only of two commercial offers which are based on microfluid science, all of which have been stopped:
- The oldest is the late Illumina NeoPrep Library System, developed and improved at its publication by Dartmouth College and is based on Advanced Liquid Logic. Nonetheless, this technology is a longtime perfect representative of renowned DMFB systems but is mainly mentioned in papers for researchers in digital microfluidics. The DMFB for the viewing is Baebies SEEKER and reproduction of infant lysosomal conditions from the King University. This platform is appealing since it has been given FDA clearance for an interesting standard medical program.
- A technology still being developed is the Oxford Nanopore VolTRAX, with several prototypes published to a specific community. According to company documentation, it seems like this system will be a USB interface for the preparation of samples and connectivity with Oxford's other technology. This device's light-weight shape factor, which will allow remote analysis, is a distinctive feature.

1.3.6 Open-Source DMFB Synthesis

- The role of digital microfluidics is on the rise. There are currently two experimental frameworks and one DMFB synthesis software application.
- Bordeaux Open Drop is an easy-to-achieve technology channel for dynamic digitally based on DC, PCB, and is reasonably affordable. It has little funding at present and whether the development will proceed is unclear.
- The Wheeler DropBot has an AC actuator and includes various prescriptions describing its application in real-life circumstances (including a field trial in Kenya). Initially, costly external lab infrastructure was required to work, but the platform is constantly enhanced in version 3.0. Also available is a biochip

monitoring and configuration tool, but sophisticated automation methods such as error recovery are not implemented.

- MFStaticSim of UC Riverside is a C++ system used to synthesize actuation sequences. It incorporates many algorithms for programming, positioning, sequencing, pin mappings, and wire transmission and provides a resulting helpful culture.

1.4 Large Scale of Flow-Based Microfluidic Biochips

Microfluidic biochips based on flow (FBMBs) are used to steer flux across networks of reaction chambers using engraved microchannels, microvalves, and micropumps [3]. In comparison to optical microfluidics, fluid flows are steady. FBMBs are an efficient way to produce the research facility because they can be rapidly and efficiently assembled and made [4]. Indeed, FBMBs are the basis of the principle of sizable implementation of microfluidics (LSI).

Generally speaking, these instruments cannot be reconfigured, as the valves need to be regulated and worked correctly for the particular goal. There are efforts to develop multifunctional flow-based biochips (RFBs) to increase their sophistication and develop a technology category akin to department door arrays (FPGAs).

1.4.1 Model of Biochip

The microfluidic circulation systems, borrowed from the semiconductor industry [4], were made using silicon and glass substrates. Polydimethylsiloxane (PDMS) is generally the leading commodity due to its simplicity and relatively precise manufacturing, elasticity, and physicochemical characteristics. Biochips of PDMS are manufactured using the soft lithography multilayer process: The biochip structure is modeled using lithography and used as molds on silicon wafers. There is a thin coating of PDMS on the chips using an elastomer. The input and output points for the fluid and vibration origins are introduced.

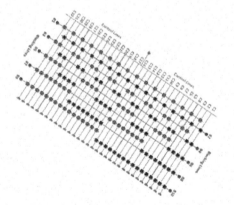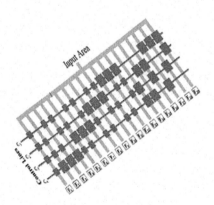

FIGURE 1.3
Generation of DMFBs.

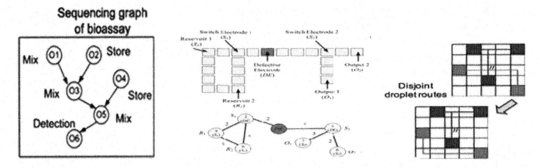

FIGURE 1.4
Structure of model chips.

Multiple levels may be configured and bundled in a dynamic channel, but only two layers are most commonly utilized, as shown in Figure 1.4.

1.4.2 Building Block of Microvalves

One of the essential building components of FMFBs is the tiny valves. The usual setup of the flow-based biochip is shown in Figure 1.4. Two layers of elastomers are network manufactured and linked together. One layer includes flow paths for sample and reagent modulation. Pumps attached at the bottom of all fluid reservoirs produce streams. A network of modulation schemes connecting to outside tension outlets is used in the second layer. Stress source triggering triggers a deflection in which the power and flow channels are cut and form a funnel. The stimulation may be conceived for regular inspiration. In standard FMFBs, the pressure operation in the control layer strengths of the valve is disrupted by the valve closing and the dynamic motion in the flow path.

1.4.3 Fully Programmable Flow of Valve Arrays

Crossing fluid valves and canals to the complete multifunction valve arrays (FPGAs) can be organized in perpendicular meshes. FPGAs allow for a full network setup. The massive microfluidic integration (MLSI) will make maximum density. Sequence analysis of the valve conditions also allows for mixing. For high-performance and parametric studies, this can be used. For example, samples of M input may be evaluated for particular n biochemical solvent interactions for the whole M by N. This kind of equipment is applied in different industries, covering surfaces test techniques and phosphodiesterase in the digital response chain. FPGAs may be improved to enable durable and sturdy systems to develop optimization targets like pin counting. Designs are also possible.

1.4.4 Protocols of Routing Crossbars

In recent times, crossbar networking has been suggested (alternatively, for routing manufacturing) so that additive manufacturing and adaptations to biochemical protocols in real time could be made. This idea has also been implemented in research on individual networks and has shown the potential to create more space-efficient structures.

The horizontal connectors are based on the two input and two output electricity grid concepts and are made utilizing a valve performance potential. The original was built in using a substratum made of ablated polycarbonate piled over polydimethylsiloxane (PDMS). The valves are shaped in the networks as displacements and may usually be opened or closed. An elastomeric membrane surrounds the valve. As the vacuum is activated (or compressed), these membranes spread into the space, and fluid passes through (or be blocked). One output port chooses a 2-to-1 multiplexer among two input ports and preferably a 1-to-2 DE multiplexer. The fundamental transposer then creates ever more complex routing topologies that may be selected from an endless number of nodes.

1.4.5 Commercialization of Microfluidic Technologies

Flow-based microfluids are among the most evolved and valuable chemicals in workbench systems among this plethora of available microfluid innovations. A remarkable example is the Biomark HD Fluidigm, available on the market as of the mid-1990s and capable of handling its various types of own built-in integrated fluid circuits (IFCs). IFCs are available in the current cartridge to carry out digital PCR and the DNA sequence.

1.5 Conclusion

The basic functioning concepts and their development, design, and advancement in the commercial exploitation of digital and flow-based microfluidics are discussed in this chapter. In all these inventions, the perfect microfluidic biochip can be realized. Microfluidic biochips can be successfully implemented in design automation, which significantly reduces the development time and improves usability. This suggests the safety and trustworthiness of these systems, like the rest of this book shows.

References

1. S. Pennathur, C. Meinhart, H. Soh. How to exploit the features of microfluidics technology. *Lab Chip* Vol. 8(1). pp. (20–22). (2008).
2. C.M. Klapp. Erich, microfluidic diagnostics: Time for industry standards. *Expert Rev. Med. Devices* Vol. 6(3). pp. (211–213). (2014).
3. T.M. Squires, S.R. Quake. Microfluidics: Fluid physics at the nanoliter scale. *Rev. Mod. Phys.*Vol. 77(3). pp. (977–985). (2005).
4. G.M. Whitesides. The origins and the future of microfluidics. *Nature* Vol. 442(7101). pp. (368–373). (2006).
5. Y. Xia, J. Si, Z. Li. Fabrication techniques for microfluidic paper-based analytical devices and their applications for biological testing: A review. *Biosens. Bioelectron.* Vol. 77. pp. (774–789). (2016).
6. S.-Y. Teh, R. Lin, L.-H. Hung, A.P. Lee. Droplet microfluidics. *Lab Chip* Vol. 8(2). pp. (198–220). (2008).
7. B. Hadimioglu, R. Stearns, R. Ellson. Moving liquids with sound: The physics of acoustic droplet ejection for robust laboratory automation in life sciences. *J. Lab. Autom.* Vol. 21(1). pp. (4–18). (2016).
8. M. Boyd-Moss, S. Baratchi, M. Di Venere, K. Khoshmanesh. Self-contained microfluidic systems: A review. *Lab Chip* Vol. 16(17). pp. (3177–3192). (2016).

9. D. Grissom, C. Curtis, P. Brisk. Interpreting assays with control flow on digital microfluidic biochips. *ACM J. Emerg. Technol. Comput. Syst.* Vol. 10(3). pp. (24–36). (2014).

10. Y. Luo, K. Chakrabarty, T.-Y. Ho. Error recovery in cyber physical-digital microfluidic biochips. *IEEE Trans. Comput. Aided Des. Integer. Circuits Syst.* Vol. 32(1). pp. (59–72). (2013).

11. F. Su, K. Chakrabarty. High-level synthesis of digital microfluidic biochips. *ACM J. Emerg. Technol. Comput. Syst.* Vol. 3(4). pp. (1–10). (2008).

12. M. Alistar, P. Pop, J. Madsen. Synthesis of application-specific fault-tolerant digital microfluidic biochip architectures. *IEEE Trans. Comput. Aided Des. Integer. Circuits Syst.* Vol. 35(5). pp. (764–777). (2016).

13. F. Su, K. Chakrabarty. Module placement for fault-tolerant microfluidics-based biochips. *ACM Trans. Des. Autom. Electron. Syst.* Vol. 11(3). pp. (682–710). (2006).

14. P. Pop, M. Alistar, E. Stuart, J. Madsen. Design methodology for digital microfluidic biochips. In *Fault-Tolerant Digital Microfluidic Biochips: Compilation and Synthesis* (Springer, Cham). pp. (13–28). (2016).

15. T. Xu, K. Chakrabarty. Parallel scan-like test and multiple-defect diagnosis for digital microfluidic biochips. *IEEE Trans. Biomed. Circuits Syst.* Vol. 1(2). pp. (148–158). (2007).

16. T.A. Dinh, S. Yamashita, T.-Y. Ho, K. Chakrabarty. A general testing method for digital microfluidic biochips under physical constraints. In *Proceedings of the IEEE International Test Conference.* China. pp. (1–8). (2015).

17. C.C.-Y. Lin, Y.-W. Chang. ILP-based pin-count aware design methodology for microfluidic biochips. *IEEE Trans. Comput. Aided Des. Integer. Circuits Syst.* Vol. 29(9). pp. (1315–1327). (2010).

18. K. Hu, F. Yu, T.-Y. Ho, K. Chakrabarty. Testing of flow-based microfluidic biochips: Fault modelling, test generation, and experimental demonstration. *IEEE Trans. Comput. Aided Des. Integer. Circuits Syst.* Vol. 33(10). pp. (1463–1475). (2014).

19. S. Bhattacharjee, A. Banerjee, B.B. Bhattacharya. Sample preparation with multiple dilutions digital microfluidic biochips. *IET Comput. Digit. Tech.* Vol. 8(1). pp. (49–58). (2014).

20. S. Bhattacharjee, S. Poddar, S. Roy, J.-D. Huang, B.B. Bhattacharya. Dilution and mixing algorithms for flow-based microfluidic biochips. *IEEE Trans. Comput. Aided Des. Integer. Circuits Syst.* Vol. 36(4). pp. (614–627). (2017).

21. D. Mitra, S. Roy, S. Bhattacharjee, K. Chakrabarty, B.B. Bhattacharya. On-chip sample preparation for multiple targets using digital microfluidics. *IEEE Trans. Comput. Aided Des. Integer. Circuits Syst.* Vol. 33(8). pp. (1131–1144). (2014).

22. S. Roy, B.B. Bhattacharya, S. Ghoshal, K. Chakrabarty. Low-cost dilution engine for sample preparation in digital microfluidic biochips. In *Proceedings of the International Symposium on Electronic System Design.* India. pp. (203–207). (2012).

23. V. Ananthanarayanan, W. Thies. Biocoder: A programming language for standardizing and automating biology protocols. *J. Biol. Eng.* Vol. 4(1). pp. (1–11). (2010).

24. G. Chen, Y. Lin, J. Wang. Monitoring environmental pollutants by microchip capillary electrophoresis with electrochemical detection. *Talanta* Vol. 68(3). pp. (497–503). (2006).

25. J.C. Jokerst, J.M. Emory, C.S. Henry. Advances in microfluidics for environmental analysis. *Analyst* Vol. 137(1). pp. (24–34). (2012).

26. S. Neethirajan, I. Kobayashi, M. Nakajima, D. Wu, S. Nandagopal, F. Lin. Microfluidics for food, agriculture and biosystems industries. *Lab Chip.*Vol. 11(9). pp. (1574–1586). (2011).

27. J. Wang. Microchip devices for detecting terrorist weapons. *Anal. Chim. Acta.* Vol. 1. pp. (3–10). (2004).

28. K.S. Kaswan, L. Gaur, J.S. Dhatterwal, R. Kumar. "AI-based natural language processing for meaningful generation information Electronic Health Record (EHR) data" accepted in Taylor Francis, CRC Press for "advanced artificial intelligence techniques and its applications in bioinformatics". ISBN No. 9781003126164. (2021).

29. R.B. Fair. Digital microfluidics: Is a true lab-on-a-chip possible? *Microfluid. Nanofluid.* Vol. 3(3). pp. (245–281). (2007).

30. M. Pollack, A. Shenderov, R. Fair. Electrowetting-based actuation of droplets for integrated microfluidics. *Lab Chip* Vol. 2(2). pp. (96–101). (2002).

31. H.-H. Shen, S.-K. Fan, C.-J. Kim, D.-J. Yao. EWOD microfluidic systems for biomedical applications. *Microfluid. Nanofluid.* Vol. 16(5). pp. (965–987). (2014).
32. K. Bazargan, R. Kastner, M. Sarrafzadeh. Fast template placement for reconfigurable computing systems. *IEEE Des. Test Comput.* Vol. 17(1). pp. (68–83). (2000).
33. D. Grissom, P. Brisk. Fast online synthesis of generally programmable digital microfluidic biochips. In *Proceedings of the IEEE/ACM/IFIP International Conference on Hardware/Software Codesign and System Synthesis.* UK. pp. (413–422). (2012).
34. D.T. Grissom, P. Brisk. Fast online synthesis of digital microfluidic biochips. *IEEE Trans.Comput. Aided Des. Integer. Circuits Syst.* Vol. 33(3). pp. (356–369). (2014).
35. J.S. Dhatterwal, K.S. Kaswan, Preety. "Intelligent agent-based case base reasoning systems build knowledge representation in Covid-19 analysis of recovery infectious patients" in a book entitled "application of AI in COVID 19" Accepted in Springer series: *Medical Virology*: From Pathogenesis to Disease Control. ISBN No. 978-981-15-7317-0 (e-Book), 978-981-15-7316-3 (Hard Book). DOI: 10.1007/978-981-15-7317-0. (2020).
36. T.-W. Huang, T.-Y. Ho. A fast reputability- and performance-driven droplet routing algorithm for digital microfluidic biochips. In *Proceedings of the IEEE International Conference on Computer Design.* India. pp. (445–450). (2009).
37. K. Chakrabarty. Design automation and test solutions for digital microfluidic biochips. *IEEE Trans. Circuits Syst.* Vol. 57(1). pp. (4–17). (2010).
38. T. Xu, K. Chakrabarty. Integrated droplet routing and defect tolerance in the synthesis of digital microfluidic biochips. *ACM J. Emerg. Technol. Comput. Syst.* Vol. 4(3). pp. (11–25). (2008).
39. P.-H. Yuh, C.-L. Yang, Y.-W. Chang. Bioroute: A network-flow-based routing algorithm for the synthesis of digital microfluidic biochips. *IEEE Trans. Comput. Aided Des. Integer. Circuits Syst.* Vol. 27(11). pp. (1928–1941). (2008).
40. H. Singh, K.S. Kaswan. Clinical decision support systems for heart disease using data mining approach. *IJCSSE* Vol. 5(2). pp. (19–23). (2016).

2

Reliability Driven and Dynamic Re-synthesis of Error Recovery in Cyber-Physical Biochips

Jagjit Singh Dhatterwal, Anupam Baliyan, and Om Prakash

CONTENTS

2.1 Introduction

In this segment, we introduce a "physical-aware" device restructuring approach that uses sensor data at intermediary crossings to dynamically rearrange the biochip, based on recent developments in digital microfluidics cybernetic implants in the application of sensor networks [1]. The differences in electrode action are recalled with the computerized maintenance management re-synthesis process, which produces new results in module placement, droplet navigation pathway, and global markets through minimal effect on the time-to-response.

DOI: 10.1201/9781003220664-2

The following are the chapter's most meaningful achievements:

- A digital biochip of a microfluid charge-coupled component (CCD)-based sensing mechanism.
- A droplet estimation and monitoring scanning algorithm based on the available information from a CCD sensor [2].
- An error-recovery technique based on durability.
- Maintainability synthesis using path redundancy based security algorithm (PRSA) and greedy strategies are paralleled strategies referring to the plans of virtual vacuum distillation.
- Three symbolic antibacterial activity simulation findings.

2.1.1 Related Work

Digital microfluidics' simplification of operating system configuration and monitoring sparked exploration into different areas of automatic semiconductor manufacturing and microprocessor implementations. A variety of techniques for architecture and design propagation [3], the placement of the device, and the drop of water [1–4] are published. A method, on the other hand, neglects the functional complexities and domain-specific limitations that come with attempting to perform biological functions and microfluidic operational activities on an electronic chip.

Predictive modeling and precision monitoring are difficult to achieve due to the dynamic and random numerous contributions that are common in biological/chemical reactions [5, 6]. Various errors can occur during the operation of a bacterial genome, in addition to manufactured flaws. For example, DNA fouling can cause several electrodes in a biochip to malfunction, and an improper acceleration voltage added to an electrode can cause electrode degradation and charged trap [7–9]. These flaws are difficult to diagnose a priori, but they do exist in some bioassays [9]. Despite this intrinsic variability, many scientific trials, such as therapeutic antibodies and drug production, necessitate highly reliable and detailed fluid-handling functions. For the concentration and volume of raindrops, each phase in a biochemical experiment procedure has an "acknowledgment range." The pH of the solution used in preparing plasmid DNA materials, for example, should be much less than 8.0 to prevent a major decrease in the lysozyme's performance [10]. If an accidental mistake happens during the procedure or the antibacterial activity protocol's conditions are not followed, the results of the whole procedure will be erroneous. To fix the mistake, all of the measures of the procedure must be replicated [11, 12]. Experimentation is repeated again and again, wasting costly reagents and difficult-to-find specimens. The following issues arise as a result of the repeated implementation of on-chip laboratory experiments:

i. Improvements to the antibacterial activity outcome that are detrimental to legitimate detection and easy implementation;
ii. Waste of hard-to-get or store materials and expensive ingredients.

As a result, strategies for tracking assay consequences on subcortical structures must be established, as well as an effective error-recovery mechanism. In the literature, error recovery in electronic microfluidics has gotten a lot of coverage. The study in [11] is the only research that has been published on the microfluidic biochip, which suggested intermediary phase tracking and rolling back error correction. The study in [11] uses a sensing device to check the accuracy of necessary and proper clause droplets at different stages of the on-chip procedure. When a sensor detects a malfunction, the resulting droplet is eliminated if the droplet's density or strength is below or above the measured distance necessary. The procedure is resumed when an investigation's production does not fulfill the consistency specifications resulting from sensor calibration. To substitute the under-qualified droplet, new product droplets will be formed.

The original sequence diagram of a bacterial genome that works as a sensor evaluates the outputs of each administering, combining, and separating process. The machine will re-execute the related administering and combining procedures if a mistake occurs at operation. It displays the latest error-recovery sequencing table, which includes the new material particles manufactured in activities 12, 13, and 14. In [11] there are some vulner-abilities in the exclusion of the detection and correction method "physical-aware" moni-toring systems:

- Overgeneralization of fault identification, as well as the stereotypes that go with it, is the first disadvantage. For the calibration of each detection process, a stan-dardized "perceived outcome" is unreliable. It is significant to mention that as antimicrobial activity behavior continues, the structure of transitional raindrops differs significantly. As a consequence, many occasions and gradually the calcula-tion are needed.

- All recovery operations are separately performed in [11]. When an error is found, all other existing bioassay-related fluidic procedures are halted. Long waiting times implemented by recovery operations can result in sample deterioration and incorrect assay results [13]. Some procedures, such as different spectrofluoromet-ric enzymes, necessitate specific durations as defined by the intervention protocol, and they cannot be extended without adding unpredictable nature into the experi-mental outcome [14].

- Where many errors occur throughout a bacterial genome, the error-recovery solu-tion in [11] is ineffective. For example, [11] indicates that all methods of data trans-mission will work and that mistakes throughout restoration don't really happen.

- The technique of error recovery in [11] ignores the problem of consistency. The accu-mulation of positively charged electrodes is normally the source of errors like the creation of droplets with irregular volumes [7, 8]. If such electrodes are used in the future, they are likely to introduce further errors [7, 8]. As a result, we would reduce the use of these electrodes, ensuring the biochips are accurate and reliable. In order to overcome the above drawbacks, we are using gradual "cyber physical" measures to develop sensor input and to monitor equipment biochip processes. We are present-ing a technique of "physically conscious" system redesign using visual data at inter-faces to restructure the biochip functionally [15] by leveraging recent developments in the implementation of wearable sensors in a digital microfluidics biochip [16].

2.2 What Are Cyber-Physical Biochips

The cyber-physical framework on biochips and on all parts of the machine is presented in this chapter. "Physically aware" control software is now possible thanks to the availability of biochip sensing systems. The term "physical-aware" refers to the fact that the device

FIGURE 2.1
Framework of a cyber-physical digital microfluidic system.

will obtain input from the sensing mechanism about the result (error-free/erroneous) of fluid-handling procedures. The monitoring system will correctly modify the microfluidic biochip based on sensor input. The different phases of the bioassay are then carried out based on substrate results and real-time reporting. Figure 2.1 shows each aspect of a cyber-physical network on the microfluidic basis. A feedback signal is sent from the monitoring system to the microfluidic biochip and a sensor on the chip tracks the effects of liquid functioning practices. The results are related to the "expected values," or pre-determined cutoffs. If the reference findings show a warning message saying that the monitoring system gets a "repeating command" and that it can execute again the following operation in which the problem occurs. This is the way to fix the error.

2.2.1 Sensing Systems of Digital Microfluidic Biochips

On optical microfluidic biochips, sensation procedures could be used in the cyber-physical framework.

The first sensing scheme is CCD camera based. CCD devices are used to independently capture the edges of raindrops, as shown in Figure 2.1.

Droplets can be dynamically identified by the control program using photographs taken by the CCD sensor. The technique for searching for droplets automatically can be represented as a "template matching" problem. The illustration of a "typical" droplet may be used to depict a pattern in this case. We transfer the biometric template to all possible locations in the image of the whole sequence and during the matching step and crop a sub-image that is the same width as the reference picture. The correlation index, which calculates the resemblance between both the prototype and the "cropped picture," is then calculated by the control program. On a pixel-by-pixel analysis, the correlation factor is determined, as seen in Figure 2.3a. Both photographs in the control program are processed in grayscale. Matrixes or vectors may be used to decode these grayscale files. Assume the template image is stored in a one-dimensional array: x D.x1;x2;:::xN/x D.x1;x2;:::xN/x D.x1;x2;:::x. Here, xi denotes a pixel's gravel level, and N denotes the template image's total number of pixels. In the same way, the cropped sub-image to be contrasted with the prototype picture can be translated as y D.y1;y2;::::;yN/. As a result, the similarity factor between these two pictures is:

$$cor = \frac{\sum_{i=1}^{N}\left(x_i - \overline{x}\right) \cdot \left(y_i - \overline{y}\right)}{\sqrt{\sum_{i=1}^{N}\left(x_i - \overline{x}\right)^2 \cdot \sum_{i=1}^{N}\left(y_i - \overline{y}\right)^2}},$$

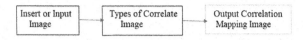

FIGURE 2.2
Structure of the template matching image.

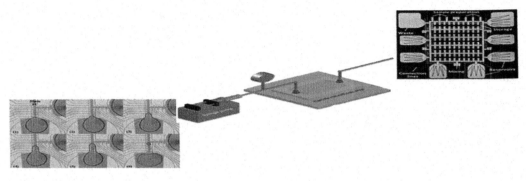

FIGURE 2.3
Structure of the microfluidic coplanar digital chip.

where xN and yN are the prototype image's and clipped sub-image's average gray levels, respectively. A similar procedure cor has a spectrum of values around 1 and C1. A higher absolute value of a correlation factor indicates a greater relation between different pictures, according to the concept of correlation.

The relationship diagram between both the prototype and the initial input image is obtained after deriving considerations of differentiation for all potential roles of absolute biosensor composition. Assume the biochip has droplets. Apposition of droplets can be calculated by looking at the correlation chart with the highest correlation variables. Figure 2.2[17] shows an example.

The original input signal of the entire chip and the pattern image with the best corresponding positions, i.e., the droplet coordinations derived from the control program, as (77, 107), (77, 147), (77, 147), and (76, 208), respectively. As a result, the control program locates the droplets on its own. The dimensions and wavelengths of raindrops can be studied further using the picture. After analyzing the image captured by the CCD sensor, the amounts and distributions of droplets can be obtained in this method. Instead of scanning the entire picture for droplets to see if the drops were moved to their correct positions, we use data modeling.

The following procedures are used to carry out this procedure:

Before beginning the experiment, several calibrations are performed. We select a wide number of sub-images both with and without raindrops and calculate their n design relationship. We find an appropriate proportion for the correlation index (Cth) based on t' estimation results: if the similarity is greater than Cth, we infer that the clipped threar' a goblet; therefore, the short and mid would not have a raindrop. We trimmed the sh mid, closed the projected places of the particles, and quantified the appropriate a' indexes while the antibacterial activity works to determine the absence/presenc'

(i) Identify exactly where errors are

(ii) Perform spot corrections as soon as they appear when using a C'
sensing system.

A downside of this strategy is that the monitoring of the cyber-physical environment requires the use of other equipment, such as CCD cameras.

The second detecting method, as suggested in [14, 16], relies on optical transmission instruments. The composition of an intermediate in the production of a digital microfluidic biochip can be calculated by analyzing the intensity of the substance in the raindrops using fluorophores [14, 16]. Whenever a raindrop comes in contact with a fluorescence indicator, different substance quantities result in different spectrums of light being emitted (i.e., different colors). Capacitive properties that transform the emitted light into electrical voltage and current signal will sense this color difference [16]. On the microfluidic series, combined properties of nanoparticles have recently been implemented [14, 16]. In [14], for example, a digital microfluidic array was combined with an optical detection system. A light-emitting diode (LED) and a photodiode that serves as a light-to-voltage converter make up this device. The concentration of compounds can be measured using the photodiode's output voltage. Thin film In GaAs photoconductor, for example, can be attached to a glass framework AF-coated coating and put into the microelectronic virtual environment. Figure 2.3 [14] shows a coplanar optical microfluidic chip with an embedded In GaAs phototransistor.

The downside of the optical transmission sensor device is that the sensor is not accurately identified in the unfortunate case, despite the fact that no instruments with a large footprint and precise orientation are needed in this process. If a transmitter is sent a drop of water from the mission to meet the specifications of the bacterial genome, we know that an error has occurred, but we cannot indicate when or where this error occurred. The CCD camera-based sensing contrast with sensor-based sensing in Table 2.1 reveals.

2.2.2 Digital Chip Software

As on-chip measurements are accessible, autonomous microfluidic biochips are enabled to use sensor information to detect error during intermediate examination inaccuracies during phylogenetic analysis execution. For a microfluidic biochip, the study in [11] suggested intermediary phase testing and rollback error recovery. The main concept behind this research is to use the on-chip sensing device to check the correctness of production raindrops at different stages of the on-chip procedure. The following is how error recovery is conducted with this method. Where a checkpoint detects an error, procedures whose

TABLE 2.1

Comparative Analysis of the Camera Sensing System CCD and the Sensing System Dependent on Detectors

	CCD camera-based scheme	CCD camera-based scheme
Accuracy of locating electrodes with defect	High	Low
Response time	Error recovery can be triggered immediately when an error occurred	Error recovery can only of the faulty operation be triggered at the end
Application for photosensitive samples/reagents	Cannot be used	Can be used
Cost	The price of the CCD camera is around $3,000	The cost of fluorescent labels is around $30/mol and the laser is around $200

outputs failed to fulfill quality specifications due to sensor configuration are re-executed to correct the problem. To allow recovery, the discarded intermediary substance droplets must be deposited in specifically specified positions on the chips. For error recovery, supplementary droplets of substances and materials must be distributed from reservoirs.

The technique for dependability error recovery as well as the algorithm for developmental effects of error recovery will be implemented based on the error-recovery process suggested in [11].

2.2.3 Biochip Software

The cyber-physical connection here between the microfluidic platform's control operating systems is then defined. For cyber-physical coupling, two frameworks are required. The CPU receives the data transfer from the first equipment when the investigation is complete, and uses that information to decode the sensor values. The second interface converts the feedback information of the control program to voltage signals which can be implemented on the electrodes of the biochip.

The cyber-physical structure is depicted schematically in Figure 2.4. A machine, a peripheral circuit, a separate microcontroller, and a biochip are all part of the device. The program on the machine sends a command signal to the biochip during the implementation of a bacterial genome. Via the solitary microcontroller, the software simultaneously provides input from the biochip's sensing mechanism. Based on sensor inputs, the control program recalculates the timetable of fluidic activities, module positioning, and droplet waterways. The key idea is to have an on-chip sensor to coordinate experimental regulation related to restructuring. When the on-chip system detects an error, the irregular droplet is immediately discarded, preventing experimental error.

Cyber-physical microfluidics' closed-loop incorporation can also be used to monitor the time it takes for biosensors to complete. Serum samples must be thoroughly combined with an enzyme-mediated reagent when evaluating glucose in blood, for instance [18]. A droplet's state is tracked by an imaging system and during the mixing process. Through studying and contrasting photographs of the droplet, the degree to which blending has

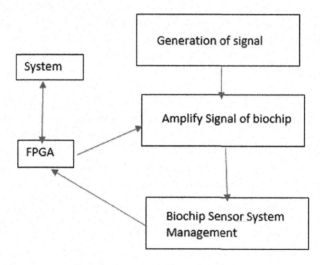

FIGURE 2.4
Flow of biochip software.

been achieved can be determined. As a result, the biochip's control program would cause it to keep mixing until the input data indicates that the droplets are properly blended. As a result, even without understanding the exact mixer execution time ahead, the combining procedure can be accurately regulated, and the on-chip glucose concentration results obtained can be as accurate as those obtained by a traditional materials scientist bench-top analysis tool [18].

2.3 Ease of Error Recovery

2.3.1 Planning of Error Recovery

The gelation temperature error recovery is formulated in this section. For each discharging, combining, dilution, and separating operation in the given antibacterial activity procedure, we use the on-chip sensing device to assess the consistency of the output droplets. It's worth noting that the period idea of upgrading detector operations is insignificant, as on-chip sensor reaction times are measured in milliseconds [19].

There are two types of fluid-handling procedures on a microfluidic biochip: irreversible and nature of intelligence operations. Microvascular procedures include blending and dilution, while reversible operations include discharge and separating. Errors that arise during irreversible procedures have relatively straightforward recovery mechanisms. If two droplets with inconsistent quantities are formed during a splitting process, the biochip will first combine the two irregular droplets into a larger one instead of separating the larger droplet. When a distributing error occurs, the chip will return the irregular droplet and will allocate a further raindrop to the proper reservoirs. There is therefore a fast recovery period and no additional droplets are required for problems that crop up throughout the process of turning things around.

Microvascular procedures have a more complex error-recovery mechanism. We will require input raindrops from procedures whose interfaces provide an unsuccessful interface activity in effort to properly relate the microvascular operations and correct the error. As a result, all predecessors of the incorrect process will need to be re-executed. If an error happens at procedure 7, for example, operations 1–6 will need to be re-run. As a result, the time expense of performing error-recovery procedures can be quite high. In order to reduce the risk of a worst-case scenario, we use the following steps:

- If only one of the separating operation's output particles the other (superfluous) drop of water was used as a reference for its eventual successor's replacement in case of a mistake at a later time. Operation, for example, is a separating operation that produces two output droplets. The fluid management of each cycle is used in the series diagram procedure, and only one of these two droplets is used as the input of installation. In the sequencing table, the discarded droplets are not visible.

- All discharge procedures are planned for implementation as quickly as practical and their output particles are deposited on the biochip if an error occurs. We also leave any droplets on the table as a buffer for any future error-recovery procedures. Those discarded contingency droplets will be returned to their accompanying reservoirs after performing the assay.

When a renal artery stenosis operation fails, the control program first scans to see if contingency raindrops stored on a biochip will sustain the operations of the organization. If the reaction is yes, then the time expense of the organization will be reduced. Anything other than that during error repair, further operations will be performed. According to the number of tests and droplet overconsumption in their error-recovery methods, the activities in the bioassay can be separated into three groups based on the above analysis. The below are the formal classifications for fluidic-handling operations:

Both irreversible operations fall into this category. If a mistake happens, they will easily be re-executed.

Category II operations are those that are nonreversible and with which basic understanding may have contingency droplets.

Category III: This takes a couple of nonreversible procedures for which the immediately responsible forerunners are unable to supply contingency droplets.

Each node in a provided decoding diagram indicates an operation. The in-number of a procedure the quantity of raindrops and the quantity of particles is specified a procedure is defined as the number of output particles. Any activity optk can be classified based on its in-number and out-number values, as listed below:

- If optk's in-number equals zero, optk is a dispensing activity. As a result, we have optk 2 Category I.

- Optk is a slicing operation if the in-number of optk equals one and the out-number of optk equals two. As a result, we have optk 2 Category I.

- Assume optj is optk's immediate forerunner. The number of contingency raindrops at optj's production can then be measured as Boptj D ONjMNj, where ONj is optj's out-number and MNj is the percentage of subordinates immediately. If the reference droplet counts for optk's nearest neighbors are all greater than 0, then optk 2 Category II is appropriate; otherwise, optk 2 Category III is appropriate.

According to the categorization result for opti, the collection of its error-recovery operations, Ri, can be deduced. Since the input droplets for Category I and II procedures are retained on the chip, they can easily be re-executed if an error happens. However, the inputs to Category III operations come from the outputs of predecessor's procedures, and we have no protection for these particles. As a result, if an error happens in a Category III operation, we must not only re-run the procedure but also backtrace to its counterparts. Suppose opte is the error procedure, and that optp1 and optp2 are its immediate successors. If these basic understandings are Category I or II processes, we will re-execute optp1, optp2, and then opte for error recovery, resulting in Ri D foptp1; optp2; opteg. If optp1 and optp2's understandings regarding are not in Category I or II, we would try to expand Ri by including their nearest neighbors. This backtracing and augmentation process must be replicated before we hit predecessor procedures which provide replacement raindrops to feed the outputs of error procedures. The trace and expansion of all the above Ri set can be summed up as follows. We define predatory to beginning with opti/ as a mapping from opti to a collection of opti's immediate predecessors in the sequencing graph.

We characterize the operator Pr as follows for a series of operations O D foptol;opto2;::::;optok g:

1: Classify opcralions into Category I, Category II, and Category III;
2: Initialization of R_i : $R_i = opt_i$:

3: Initialization of intermediate variable $Re : Re = p_r(R_i); Re : Re - P(R_i)$

4: **while** $\{Re - R_i\} \cap \{$Set of operations in Category III$\} \neq \emptyset$ do

5: Update $R_i R_i = P_r(R_i)$:

6: Update $Re Re = P_r(R_i)$;

7: **end while**

8: $R_i = Re$;

9: R_i is the set of recovery operation for opt_i;

$$\mathcal{R} : \mathcal{O} \rightarrow \bigcup_{i=o_1,o_2,\dots o_k} \{opt_i, opt_j \mid opt_j \in pred(opt_i), \forall j\}$$

The method can be used to extract the set of error-recovery procedures Ri for any operation opti. According to the foregoing section, we may extract the set of recovery efforts Ri for any process opti.

We can add edge among activities in the set Ri describing the relationship between procedures in the initial sequence graph, and thus extract the error-recovery graph GRei for opti. We can re-execute procedures in GRei for error recovery if an error happens in opti. It's worth noting that certain electrodes on the biochip have been purposefully left unused and are being used to store backup droplets. Storage cells are dedicated and reserved for all electrodes on the chip's boundary. As a result, replacement particles can be delivered on a cybernetic implant safely.

2.3.2 Ensure Reliable Error Recovery

It is unworkable that reliable processes are maintained by simply restarting the course of treatment about which an accident appears when an error is found during the implementation of a bacterial genome. This is because errors in sample preparation execution are often triggered by electrode abnormalities; therefore, numerous errors can happens at different times in the very same biochip region. In the accompanying main instances, the objective of the charging phenomenon and DNA losing possession are seen.

Psychologically trapping charge and persistent voltage can cause mechanical failures as microfluidic digital biochip architectures are overly powered [7, 8]. Charging is a state where the charging is concentrated in the superconducting insulating material of the biochip. The load is stuck and decreases the separation technique strength and antimicrobial activity equipment failures.

- Let's pretend that Electrode 1 has stuck charge in its dielectric insulator plate, but Electrodes 2 and 3 don't. Excitation current and voltage are connected to Electrode 1 and Electrode 3 in order to perform a splitting process. The electrowetting power would be weakened by the voltage trapping on Electrode 1. Inequitable forces will break the raindrop, and the two raindrops that arise can have different amounts. Additional errors can occur if we merely re-execute the separating process while continuing to use Electrode 1. Worse, the charge trapping mechanism has the potential to permanently degrade the electrode's dielectric properties [7, 8]. As a result, if an error is observed, the electrode at which the problem happened must no longer be used to introduce fluid-handling operations in order to maintain the biochip's reliability.

- Droplets comprising macromolecules (such as DNA) have the potential to foul the electrode surfaces [20]. As a consequence, the distribution of droplets can shift in unfavorable ways. Other droplets can become infected while these infected interfaces manage to be used. The region of the decontamination of DNA is a member of the mixture combining operation's output droplets which could have odd concentration.

- To achieve maintainability error recovery, we employ a basic technique. Where an error is observed, the bioassay's execution is modified as follows:

  - The protocol with the fault is restarted. In future procedures, the instruments that can lead to errors are not used. The servo controller monitors what more storage is used by each operation. The field where the mistake happens will then be returned, based on the error raindrop. Both electrodes in this area are considered to be suspect for defects. As a result, these electrodes would be circumvented.

2.3.3 Difference among Numbers of Sensing Schemes

In terms of fault detection, error repair, and complex re-synthesis, the individual handling of strategies presented.

The evaluation of a stuck charge on an electrode will be used to demonstrate the differences in these two sensing systems. Assume a droplet separating procedure happens for inconsistent particles. Electrode 1 can accurately be defined as the electrodes with remaining charges in the CCD camera-based sensing device since the droplet density is smaller than the implementation agency.

In the other side, the outputs of the connector, which comprises of Electrodes 1, 2, and 3, will no longer be used by the optical detector-based sensing device. Electrodes 2 and 3 will no longer be used, according to the diagnostic outcome of the CCD camera-based sensing device, resulting in a waste of on-chip energy.

The bioassay is then used to demonstrate the variations between the two sensing systems.

Assume that the droplets for distributing operations come from various distributing ports. For both of the bioassay's combining processes, as well as their production. The product of module positioning in relation to synthesis. It's worth noting that "resource" refers to an electrode array component that is occupied by the blended phase. A converter is located at the top right angle of the container by the configuration of the electrodes, e.g., the back corner of the combined module M1 seems to be in the sixth and second rows and has a two-row, three-column electrode array. As a result, the mixer is referred to as a 3 2 mixer at the venue (2, 6).

Assume that the DNA-fouling anomaly happens after the operation has been running for 3s in Mix 3. Only after Mix 3 has been done is the quality of Mix 3 tested for the electronic detection system. As a result, the detection and correction procedures are launched at the time instant t D 10. The CCD camera-based device detection and correction process will start seriously once DNA contamination occurs at clearing t D 3. These are the summaries of the three instances. Recovery can only be activated at the conclusion of an incorrect activity in a detector-based sensing device, according to our findings. Recovery can be activated directly after a mistake happens in sensing device on the CCD camera. In the sensing on the CCD camera device, on the other hand, restoration can only be initiated after an error has occurred.

It's worth noting that the camera's light will affect certain biochemical substances in the droplet, such as fluorescent markers [21]. As a result, we must select a detector-based digital dashboard to track experiments with photographic film packages.

2.4 Physical-Aware Sensor Software

With hardware that can give input to the monitoring systems now available, it's time to create technology that can understand sensor information and adjust the output of the synthesizing efficiently using flow-through organizational upgrades. This implies, among these modifications are the resource connections, the position of the component, and the droplet navigation.

The re-synthesis process is divided into two phases: the first is off-line data planning prior to antibacterial activity execution, and the second is online control for fluid-handling activities as well as dynamic antibacterial activity re-synthesis. The following sections provide more information.

2.4.1 Offline Data Preparation

Order to convert the bioassay's sequencing graph to a guided acyclic graph (DAG) and saving the DAG in memory for use by the control program is the first step in data planning. The vertices in this DAG reflect microfluidic data aggregation, while the edges display precedence relationships between them. The ancestors and descendants of any procedure can be found by doing a depth-first scan on the graph [22].

Identifying error limits for each activity is the second stage of data processing. The precision required for the bacterial genome determines these standards, which are stored in the system as a table for use by the management system. When during the immunoassay performance, a detecting answer is over the spectrum of predesignated thresholds, we conclude that the next operation has produced an error.

Deriving bioassay's initial synthesis outcome is the final step in data processing. The sequence chart of the quarterly sales and on-chip properties are mapped to the preparation, resource links, placement module, and the outcomes of droplets navigation for every protocol operation.

For a sequence network of n processes, the conclusion may be expressed in terms.

$$ReRe = P_r(R_i)$$

$$\mathscr{S} = \{M_{opt_1}^*, M_{opt_2}^*, ... M_{opt_n}^*\}$$

where $M_{opt_i}^*, 1 \le i \le n$ is the synthesis output for the ith operation opt_i. The element $M_{opt_i}^*$ can be viewed as an ordered 6-tuple:

$$M_{opt_i}^* = < ts(opt_i), te(opt_i), x(opt_i), col(opt_i), row(opt_i) >$$

where ts and te are the organization's start and stop times, including both; x and y are the service module x- and y-coordinates; and col (row) is the number of columns (rows) represented by the process in the sequence.

Several suggested algorithms for solving this optimization technique have previously been released for digital microfluidic biochip computer-aided design and development. For instance, PRSA-based propagating techniques may be utilized to quickly provide optimum outcomes of synthesizing [23].

Because after optimum summary findings are received, another off-data acquisition stage is done. Antioxidant activity is then carried out based on the initial propagation result, followed by droplet monitoring in real time.

2.4.2 Online Monitoring Microfluidics Chip

The control program shall take the following actions during the execution of the bacterial genome.

2.4.2.1 Step 1: Detecting of Error

The procedures of error detection for the intelligent control CCD-based camera monitoring program and technology vary. Outputs of each activity shall be submitted to an on-chip detector for the detector-based sensing systems. Following each operation of the optical detector, the program contrasts the detection effect with the preset error thresholds. If we do not satisfy the condition of the experiment, the optical detection result concludes that a mistake has occurred. Error detection triggers the detection and correction procedure: i.e., the program adjusts the synthesis effects dynamically to restart the process where the problem happened. The program also circumvents all questionable electrodes where the mistake happened. For example, presume that the process performance droplet doesn't fulfill the criteria and that the defect can then be concluded in the resources devoted to opt. This area will be circumvented in order to guarantee the stability of future operations.

The program will watch all droplets in the biochip in real time for the CCD camera-based sensing device. The CCD camera simultaneously detects the colors and thicknesses of the particles and evaluates them for comparisons. This triggers error repair as soon as a mistake happens.

2.4.2.2 Step 2: Updating Structure of Graph

The control system specifies the recovery operations necessary when an error occurs. The program adjusts the sequencing of the bioassay according the type of operation if there is a mistake in the course of the operation. If a permanent error happens, the recovery process is simple. In the event of a mistake during a not phase change, before activities can be identified, the programming language has to verify the prior procedures and give backing nanoparticles to send restoration outputs to the saved method. The pseudo-code for sequence graph modification. The GRei define for operational opti for the detection and correction diagram. The revised sequencing diagrams, respectively, are the defects in procedures in Categories I, II, and III. You will find the classification of operations. Importantly, the recovery subprocesses might vary according to the error for certain operations. For example, two droplets are generated; one of them is therefore implanted on the semiconductor as a "duplicate droplet" for image transmission. Where a single mistake occurs, the biochip re-runs operations. If an error happens in an operating ancestor, however, the recuperation subroutine is different for operation 14 if an error occurs, for instance, it is used as an input to restore error by using the operating backup droplet. When another fault happens subsequently at service, the operating output contains no extra droplets. Thus, the operational recovery simulation environment must be extended and activities are now included. Therefore, where a fault happens in the process, the repair steps are entirely different.

Upon definition of an operation retrieval stored procedure, the control program updates the graph sequence and the accompanying DAG, which then performs the developmental effects process.

Steps of Graph Sequences

1: Derive the graph $G_{original}$ by deleting edges between the erroneous operation and its immediate predecessors in original sequencing graph;

2: Derive the error-recovery graph $^{G}_{Rei}$ for $^{opt}_i$;

3: Copy $^{G}_{Rei}$ and label the nodes with different names;

4: Derive the union graph for $^{G}_{Rei}$ and $^{G}_{Original}$.

2.4.2.3 Step 3: Concept of Proper Handling

When a bug is found at a control point, a different visualization of the subjected to intense (which include careful treatment of intermediate results) of the phylogenetic analysis would be triggered in the cyber-physical framework envisaged. This method is called a re-synthesis based on the original design derived from the a priori neutron activation. The re-synthesis method comprises the following:

- Other processes must be prevented from interrupting. Take the example below. For bioassay with transcription, an error-recovery procedure at particular instant 10 should be activated. The Mix 1 procedure is introduced when the error-recovery mechanism is initiated. In order to prevent disruption, the results of the re-synthesis shall be the same in the previous studies of the synthesis.

- In the findings of the re-synthetic analysis, electrodes at which the mistake was, it is sometimes considered that it should be circumvented.

We suggest two re-synthesis methods for complex re-synthesis to fulfill these criteria. First, a local greedy algorithm, and secondly, a multi-objective optimization algorithm based on PRSA [3]. The first step is determining all operations for the greedy algorithm that has to conform to the outcome of re-synthesis. These include: the error-recovery diagram procedures, incorrect procedures, and corresponding operational processes to be carried out on electrodes with defects in the original propagation result. The original product of the synthesis will be used for further operations.

As a module positioning with challenges challenge, dynamic recon synthesis on the microfluidic array can be designed as formulation outcomes for certain procedures are defined. The procedures that are carried out a priori as "obstacles" are here determined by the initial synthesis outcome, and other surgeries needed for regeneration are re-synthesized and inserted greedily in the remaining biochip region accessible. The following are thorough measures. The following.

First, the control program puts all activities that will need to be replanted in a list of priorities, depending on the quantum mechanical sort outcome for the procedures. These procedures include error repair operations and all error descendants. For each process on the queue, the program then gives priority. The lowest priority is given to the stored procedure operations, i.e., the complete competence at the topographical level, whereas the assignment of scarce in the queue is the "bassiest" operation (top of the topological sort).

Then on-chip resources are allocated to these procedures in the control program. R varies with the time t in the on-chip resource package. In order for the current process to be the primary concern, the control program would look for appropriate tools. For example, if a mixing procedure is the process with the top importance, the machine will look for an accessible mn subarray of the electrode that has not been controlled from time to t C 4t. Time to t here 4t is the time necessary for activation of the mn subset of sensors. If sufficient vacant opportunities are distributed, resource coupling will be successful, and the new schedule will be taken into account. Instead, the surgery will be deferred for supplies. If the multi-resource possibilities are delivered concurrently, the software will choose one and link it to a matching activity. The procedure will be withdrawn from the priority queue after the commodity is bonded and the start/stop time determined. Be aware that the above measures can also be used to produce re-synthesis results where many errors are observed at the same time. In this case, several recovered procedures are enabled simultaneously and a priority queue for every rehabilitation process is created by the control program. After these priority queues have been combined, the control program gives a topologically dependent priority to each feature. Finally, with any combined priority queue process, the control program calculates different synthesizing performance.

There are challenges in finding appropriate instruments for use in a microfluidic biochip using a computer system with just a limited number of MN and P terminals. This algorithm of re-synthesis is "O" (i.e., "The whole vacant rectangles") (MN C P). Each electrode surface port in the array is scanned and monitored extensively by software. Since there might be a purpose in optimizing the amount of distribution channels and the techniques for big arrays, the worse condition is O (MN). Computing cost is O in some sections of the algorithm (1). The completion of the re-synthesis method is thus O (MN).

1: Locate the error operation at control locations based on feedback;

2: Identify the procedures to be altered and put in the priority queue Q;

3: Remove all initial operating synthesis results;4: while Q = 0 / do

5: Searching for the resource accessible for operation q0 that has the greatest priority in Q;

It displays the scheduling that matches the graph scheduling and both display the scheduling that matches the graph numbering. We present only the timetable for combining procedures in order to be clear. This is the timeline obtained with the algorithm of [11] error recovery.

We can observe that Mixture 1 has been paused ten times during data transmission. Around 18-time places and 28-time places raise the minimum traditional statistics which could be recognized for only certain conditions. The related complex planning outcome is if the error is found in Mix 3 at 10; the current Mix 1 process will be performed on the basis of the initial outcome of propagation. It is expected that the computing benefit of implementing the previous phone is 1 time slot. The calculating time is really smaller than the liquid organizational processes reduction in time at least in an astronomical amount. At the latest date 11, a re-synthesis result will be produced based on the revised Mix 1, and 12 will be completed without interruption. At time 18, the experiment is completed. Throughout these bioassays, no time penalty or obstruction of other activities is executed "seamlessly."

A PRSA-based optimization algorithm approach from [3] is also used to solve the re-synthesis problem. The inputs and limitations of the issue with re-synthesis vary from the original problem with synthesis. Assume that the re-synthesis dilemma operation set is P0 and the limitations set is C 0. P0 and C 0 can be derived based on the added P and C.

On the set P, we identify first a constructor T. T is a projection from the set to the sequence of steps currently begun instantly at t.

T .t/ W P ! P.t/ D fopt$_j$jM$_{opt}$i .1/ tg

In case an error in operation opti is found at the moment, P0 D P[Ri [OQ P.t/] can be written in a series of operations which require synthesis. In this respect, Ri is the set of recovery operations that matches the wrong operational opti, and OQ is the set of subsequent operations that are carried out on electrodes with defects during the initial synthesis result. The procedure is adopted to determine the Ri operations. Then, the OQ operations can be calculated based on the module positioning details contained in the original result and on the position of electrodes with defects. For opti 2 P0 Mopt0, we have written new synthesis findings. Besides the set of limitations C, the product of the re-synthesis must meet the restriction that the area where a mistake is considered is not to be in use. The topic of re-synthesis optimization can be written as follows:

minimize: Max $_0$fM$_{opt}$^{0} i .2/g opt$_i$2P

Using PRSA-based synthesis technique implemented in [3], the above enhancement problem can be resolved. This approach can be used to extract global propagation results with fast completion of the test, whereas for a standard bacterial genome, the CPU time is in the order of 20 min [11]. This approach is also not ideal for online re-synthesis effects calculation.

2.5 Compilation of Sensing Scheme Optimization

In this part, we test the re-synthesis method in representative bioassays for data transmission that are particularly likely to be fluidic. Completion times are measured for the two sensing strategies; the m actually results from the optimization method and the global optimization algorithm centered on PRSA are also discussed.

2.5.1 Developing of Plasmid DNA

We replicate first the biosensors known as the "alkalinelysis specimen preparation of plasmid DNA with SDS" [15, 24]. A combination of three reagents is essential during the separation process. The three reactants are:

1. R_1: Alkaline lysis Solution I [50mM glucose, 25mM Tris-HCl (pH 8.0), 10 mM EDTA (pH 8.0)].
2. R_2: Alkaline lysis Solution II [0.2N NaOH, 1% SDS (w/v)].
3. R_3: Alkaline lysis Solution III (5M sodium acetate, glacial acetic acid).

The required concentration of the mixture is 0.22% of R_1, 0.44% of R_2, and 0.34% of R_3, which can be approximated as $ReRe = P_r(R_i)$ of R_1, $\frac{56}{128}\frac{28}{128}$ of R_2, and $\frac{44}{128}$ of R_3.

The processing diagram in mixing R1, R2, and R3 to achieve the necessary concentrations is obtained. This bioassay is converted to an electrical array and is used to store all electrode at the array border. When mistakes are observed, the cyber physical microfluid system's error-recovery capacity may be assessed on the basis of the completion period of the bioassay. Even during bioassay, the errors are uniformly inserted into the chip and the timeframe of both sensor systems is comparable. In this case, the fulfillment time comes from the implemented optimization method. The findings are on balance ten times the obtained values by repeating the tests. For this (no error recovery), where an error happens during the bioassay, the final result of the whole procedure is wrong. In order to correct the mistake, it is necessary to remove the biochip and to replicate the experiment on a new biochip. If we conclude that the procedure is effective, the lifetime of the bacterial genome in the defect-free case will be twice as long. Based on these data, error recovery can minimize the completion period for bioassay and can also reduce the use of biochemical substances. In data transmission based on performance, in other procedures, electrodes that are considered to have a defect are not used. However, the area in which the mistake occurs will continue to be employed in following procedures during the rehabilitation process of performing problems [15], where an accident happened throughout implementation. These defective electrodes will lead to additional failures. The following approximation is developed for comparison of performance and dependability detection and correction of vital operational durations. Completely at random, we pick one choice as the first instance of the error in the implementation of the bacterial genome in the reliability-obvious error recovery. Optfe-based electrodes are alluded to as "default electrodes." We conclude that there is a Pfail risk that this experiment will fail, when another operation is performed again on these defective electrodes. We run the simulation 15 times for reliability-informed error detection for a fixed value of Pfail and detect typical execution time.

Then we choose another procedure as optfe arbitrarily and display the results again. Optfe's electronics now include an effort exerted. We found that in the event of further electrons being unreliable, the total duration for reliability-recovery errors is greater. The reliability-based issue healing process doesn't rely and preserves minimal time complexity on the source of the semiconductor failure.

2.5.2 Interpolating Protein Dissolve

We assess the transcription for three binary protein assays and error recovery. These tests lead to a test organism using two processes, namely interpolation and incremental dilution. It displays the two protocol sequencing charts [11]. The procedures are outlined in [11] for these two biomarkers.

Completion of biosensors in the case of CCD-based sensor network even without route discovery interpolation of preliminary testing. A 1010 electrode array is used in the bacterial genome.

It records the power consumption of the linear interpolation mix bioassay when many errors are added. The completion time here is specified only for fluid handling and excludes re-synthesizing time used on the CPU. We can observe that the total duration of PRSA algorithm total duration and that of the gloomy algorithm are nearly identical; however, the CPU times are different for these two algorithms. The simulation has been done on an Intel i5 6GB memory controller of 2.6GHz. Based on the same original propagation outcome, all re-synthesis algorithms are applied. The CPU time needed to compute the re-synthesis results with PRSA is about 33 minutes, which is ten times greater than the duration of the bioassay. Although the CPU time for the optimization method is less than 5 seconds, that is just 2.5% of the project duration for bioassays. The bioassays resulting from the convolutional neural network are just marginally longer for the PRSA. Notwithstanding, because of the low CPU time, the greedy algorithm is more appropriate for interactive re-synthesis [25].

Although the PRSA solution is less appealing to decision-making in real time, the gold-based algorithm offers a valuable calibration point and demonstrates that it is efficient for prompt phylogenetic analysis completion. In addition, the PRSA-based approach can be used as the basis for alternative techniques of route discovery by pre-calculating and pre-loading retrieval timelines. We equate the power consumption for the maintainability and reliability-obvious error-recovery strategies with the exponential dilution protocol implemented in [11]. First, we pick arbitrarily one optfe operation as the first instances of failure in the bioassay, where optfe is a 14 electrode subarray dilution procedure. We then initialized Pfail to reflect the likelihood that further actions on this deficient antenna structure would also fail. Then we execute the experiments 15 times according to any value of Pfail and draw the average reliability-obligeable error-recovery project duration. The faulty electrodes are, by comparison, sidestepped with the error recovery due to durability. Thus, the time duration is separate from Pfail for the serviceability error recovery. At the same time, we prevent the dilemma that any collection of faulty electrodes will lead to repeated failures, thus reducing the number of mistakes in the bioassay and improving the investigation's quality. The cost of the procedure is minimized when less reagents/samples are collected.

2.6 Conclusion

We've seen in this chapter how recent progress can be used in integrating a sensing device into a biochip for digital microfluidics to render biochips error resistant. For the reconfiguration of the "physical-aware" framework, we have introduced a cyber-physical solution, using sensor data at intermediary control points to automatically reconfigure the biochip. Real-time monitoring techniques were considered dependent on combined optical detectors and CCD cameras. In order to dynamically create new timelines, module placing, and droplet routing routes for the bacterial genome with minimal effect on response time, two separate sensor-driven re-synthesis technologies have been developed. These two methods were analyzed and compared with the time required for re-synthesis bioassays and CPUs. The synchronization of the control program of the physiological and microfluidic biochip makes it possible to use sensor data as input to decide full operations, improve electrode acceleration sequencing for subsequent processing, and reorganize the biochip dynamically. Four selected protein biosensors have shown the feasibility of the proposed solution by simulations.

References

1. T.-W. Huang, C.-H. Lin, and T.-Y. Ho. "A contamination aware droplet routing algorithm for the synthesis of digital microfluidic biochips". *IEEE Transactions on Computer-Aided Design of Integrated Circuits and Systems*. Vol. 29(11). pp. 1682–1695. (2010).
2. E. Maftei, P. Pop, and J. Madsen. "Routing-based synthesis of digital microfluidic biochips". In *Proceedings of the International Conference on Compilers, Architectures and Synthesis for Embedded Systems*. England. pp. 41–50. (2010).

3. K. Chakrabarty, and F. Su. *Digital Microfluidic Biochips: Synthesis, Testing and Reconfiguration Techniques*. Boca Raton, FL: CRC Press. (2006).

4. T.-W. Huang, and T.-Y. Ho. "A two-stage ILP-based droplet routing algorithm for pin-constrained digital microfluidic biochips". *IEEE Transactions on Computer-Aided Design of Integrated Circuits and Systems*. Vol. 30(2). pp. 215–228. (2011).

5. M. Iyengar, and M. McGuire. "Imprecise and qualitative probability in systems biology". *International Conference on Systems Biology*. (2007). pp. 1–24. DOI: 10.1007/978-3-319-56255-1_1

6. K.S. Kaswan, L. Gaur, J.S. Dhatterwal, and R. Kumar. "AI-based natural language processing for generation meaningful information Electronic Health Record (EHR) data" accepted in Taylor Francis, CRC Press for "advanced artificial intelligence techniques and its applications in bioinformatics". Taylor & Francis Group. (2021).

7. J. Verheijen, and M. Prins. "Reversible electrowetting and trapping of charge: Model and experiments". *ACS Journal Langmuir*. Vol. 15. pp. 6616–6620. (1999).

8. J. Park, S. Lee, and L. Kanga. "Fast and reliable droplet transport on single-plate electrowetting on dielectrics using nonfloating switching method". *Biomicrofluidics*. Vol. 4(2). pp. 1–8. (2010).

9. E. Welch, Y.-Y. Lin, A. Madison, and R. Fair."Picoliter DNA sequencing chemistry on an electrowetting-based digital microfluidic platform". *Biotechnology Journal*. Vol. 6(2). pp. 165–176. (2011).

10. S. Kotchoni, E. Gachomo, E. Betiku, and O. Shonukan. "A homemade kit for plasmid DNA mini-preparation". *African Journal of Biotechnology*. Vol. 2. pp. 88–90. (2003).

11. Y. Zhao, T. Xu, and K. Chakrabarty. "Integrated control-path design and error recovery in digital microfluidic lab-on-chip". *ACM JETC*. Vol. 3. pp. 11–20. (2010).

12. C. Mein, B. Barratt, M. Dunn, T. Siegmund, A. Smith, L. Esposito, S. Nutland, H. Stevens, A. Wilson, M. Phillips, N. Jarvis, S. Law, M. Arruda, and J. Todd. "Evaluation of single nucleotide polymorphism typing with invader on PCR amplicons and its automation". *Genome Research*. Vol. 10(3). pp. 330–343. (2000).

13. H. Singh, and K.S. Kaswan. "Clinical decision support systems for heart disease using data mining approach". *International Journal of Computer Science and Software Engineering*. Vol. 5(2). pp. 19–23. (2016).

14. J.S. Dhatterwal, K.S. Kaswan, and Preety. "'Intelligent agent based case base reasoning systems build knowledge representation in Covid-19 analysis of recovery infectious patients' in book entitled 'application of AI in COVID 19' accepted in Springer series: Medical virology: From pathogenesis to disease control". ISBN No. 978-981-15-7317-0. (2020). (e-Book), 978-981-15-7316-3 (Hard Book). https://doi.org/10.1007/978-981-15-7317-0.

15. R. Evans."Optical detection heterogeneously integrated with a coplanar digital microfluidic lab-on-a-chip platform". In *Proceedings IEEE Sensors Conference*. America. pp. 423–426. (2007).

16. N. Jokerst, L. Luan, S. Palit, M. Royal, S. Dhar, M. Brooke, and T. Tyler. "Progress in chip-scale photonic sensing". *IEEE Transactions on Biomedical Circuits and Systems*. Vol. 3(4). pp. 202–211. (2009).

17. Y. Luo, K. Chakrabarty, and T.-Y. Ho. "A cyberphysical synthesis approach for error recovering digital microfluidic biochips". *Proceedings Date*. Spain. pp. 1239–1244. (2012).

18. Y. Zhao, and K. Chakrabarty. "Digital microfluidic logic gates and their application to built-in self-test of lab-on-chip". *IEEE Transactions on Biomedical Circuits and Systems*. Vol. 4(4). pp. 250–262. (2010).

19. R. Fair. "Digital microfluidics: Is a true lab-on-a-chip possible?" *Microfluidics and Nanofluidics*. Vol. 3(3). pp. 245–281. (2007).

20. B. Hadwen, G. Broder, D. Morganti, A. Jacobs, C. Brown, J. Hector, Y. Kubota, and H.Morgan."Programmable large area digital microfluidic array with integrated droplet sensing for bioassays". *Lab on a Chip*. Vol. 5. pp. 3305–3313. (2012).

21. M. Jebrail, and A. Wheeler. "Let's get digital: Digitizing chemical biology with microfluidics". *Current Opinion in Chemical Biology*. Vol. 14(5). pp. 574–581. (2012).

22. U. Resch-Genger, M. Grabolle, S. Cavaliere-Jaricot, R. Nitschke, and T. Nann."Quantum dots versus organic dyes as fluorescent labels". *Nature Methods*. Vol. 3. pp. 763–775. (2008).

23. R. Sedgewick. *Algorithms in C: Graph Algorithms*. Boston, MA: Addison-Wesley. Vol. 23. pp. 25–36. (2001).

24. S. Kirkpatrick, C. Gelatt, and M. Vecchi. "Optimization by simulated annealing". *Science*, Vol. 220(4598). pp. 671–680. (1983).

25. Y.-L. Hsieh, T.-Y. Ho, and K. Chakrabarty. "A reagent-saving mixing algorithm for preparing multiple-target biochemical samples using digital microfluidics". *IEEE Transactions on Computer-Aided Design of Integrated Circuits and Systems*. Vol. 31. pp. 1656–1669. (2012).

3

Online Decision-Making-Based Cyber-Physical Optimization in PCR Biochips

Piyush Parkash, Nitin, and Shalini Singh

CONTENTS

3.1 Introduction

A significant constraint for a wide range of genomic bioanalysis is the quantity of physiology sample [1–4], namely deoxyribonucleic acid (DNA). For example, during the analysis of gene quantities of tumor DNA by functional genomic crossbreeding, the procedure takes hundreds of nanograms of DNA strands to identify the fluorescence. [2]. A significant number of DNA strands straight from organic materials is challenging to acquire. The PCR algorithm is utilized to multiply (faithfully recreate) the original DNA strands [2, 3, 5] as the first step in this bioanalysis to address this difficulty. The genomic optimization algorithm may be split into three distinct phases based on the kinds of activities involved [6].

DOI: 10.1201/9781003220664-3

One stage is the development of PCR samples. The second stage is an enhancement of DNA furrows. Following DNA amplification, a third stage involves mixing the DNA furrows with other droplets to detect the amplified furrows' absorption and hybridization [4]. The second phase consists of the amplification of the DNA furrows.

Digital microfluidic biochips (DMFBs) are widely utilized for economic evaluation and are feasible for biomolecular interactions. The systems for controlling many PCR processes on a single chip control system [3]. Droplets on a DMFB may be manipulated by employing an electrode effect [3, 7, 8], a two-dimensional electrode array. A direct measurement for the droplet and response time in the investigation may be accomplished with microfluidic system sensors like photodetectors and immunofluorescence microscopes [3, 9]. The DMFB platform offers numerous benefits in implementing PCR in comparison to older equipment and testers. The whole PCR technique may be implemented smoothly and react to reduced usage. Short-term results include quick cooling/heating rates and various integration of processing modules [5, 10]. The complete PCR system may also be decreased in size and electricity usage.

This is a perfect example. A heater creates a "hot" area and a "cooled" region. Typically, 94_1 and 65_1 are placed in the heated and chilled areas.

First, "PCR mix" is created on the biochip and is kept in a reservoir, including dNTPs and $MgCl_2$ Taq template DNA and aqueous reactions buffers. The "PCR mix" droplets are then combined with the target DNA sequences sampled. The gutter can then be mixed alternately [3, 7]. There are two stages to each thermal cycle: one to heat the droplet, dividing into two strands of the double-stranded DNA inside the droplet known as "DNA melting," and the other to cool the droplet. Every strand of the DNA recognition site for manufacturing "primary annealing" of new DNA strands. This allows for millions of copies of the original DNA strands [3]. As fluorophores mark the amplified DNA strands in the droplets, the target DNA strands may be generated by monitoring their fluorescence intensity at each PCR cycle using the assistive technologies of microfluids [4–6].

Some biochips are integrated with the detecting electrodes (DEs). [11–14] to allow on-chip hybridization of amplified DNA. The surface of these DEs is manufactured using golden sensors with various immobilized DNA samples [11, 13]. The droplets containing the DEs are left immobile in expanded nucleotide bases following the PCR process. DNA hybridization happens when complementing strands are the PCR amplification strands and the samples are immobilized on the detectors. DNA fragments can therefore be examined in the amplified DNA strands.

Despite several advantages, there are four limitations in the earlier work on PCR implementation for DMFBs:

- The function modules utilized in the PCR three-stage process are created in all previous works by themselves [3, 7], and the same design cannot be combined effectively. This inability of the three steps often influences the accuracy and effectiveness of the entire immunoassay.

- Bioanalysis is not regarded to be inherently random and complicated. When the droplet is delivered to the biochip, the droplet cannot contain enough PCR DNA (called a "vacuum droplet") [15]. The biochips built for earlier work cannot control the presence of empty droplets; therefore, the wasteful execution of the PCR cannot be stopped.

- In earlier work, there was no consideration of interference (electric, thermal, optical, fluid) with biochip devices arising because of the nearness effects. High temperatures surrounding the heater, for example, can lead to biological sample degradation in a reservoir [16]. Therefore, they cannot be positioned too near each other for the heater and the pool.

- The mixing schedule is unknown for the previously developed PCR biochips [6]. In combining procedures, this might cause excessive transport of droplets during the PCR reaction, adversely influencing the biochip's performance and speed.

It offers the first design technique, which optimizes the cyber-physical, digital biochip PCR method to resolve the short remarks mentioned previously. This work has the following significant contributions:

- A statistical model for gene delivery is utilized to detect if the droplet contains enough DNA strands to perform the PCR to solve the challenge of empty droplet detection. The luminous intensity measured by photodetectors (PDs) [17–19] or by a fluorescence microscope is retrieved by the system online in order to determine whether the heat cycles can be continued or not. This enables us to develop an effective PCR biochip for cyber-physics on a DMFB platform.
- Suggestion for developing a technique of designing, which includes reservoirs, sensors, and thermal units, taking the interference between the on-chip parts into account. A heuristic approach reduces the biochip size and electrode count and meets proximity requirements.
- The positioning of modules considers the time costs of the transit of droplets and the PCR biochip fault tolerance. The total mixing/mixing/dilution duration is adjusted for a particular bioassay, and a droplet routing mechanism is devised to circumvent electrode errors. Table 3.1 presents the timeline for the historical development of biosensors.

The visibility of outlets for cyber-physical biochips in the sensor module is explored. Mixing is planned to guarantee that the sensor system can monitor the droplets from top and side perspectives in real time.

3.2 Online Strategic Planning Cyber-Physical Biochip

The lab configuring for the PCR biochip [9] is presented. The biochip sensor system monitors the antioxidant capacity in the droplets and gives the control software with input. Thus, the program may manage PCR process execution based on the sensor data.

In PCR, fluorescence is generated when a DNA coloring qualified majority binds to double-binding DNA. Therefore, a rise in DNA during PCR results in an expansion in the fluorescence intensity [4–6]. The sensor system measures the fluorescence brightness of the droplet and supplies the control program with input. Therefore, the software may manage the PCR prouse confinement effects to manipulate the PCR process's performance number in a droplet's statistical method.

DNA droplet number is a nonlinear function that may be seen in the collection and DMFB processes [15, 20]. We find that the distribution process quantitatively follows Poisson from various information obtained from biochemistry studies where there is little

TABLE 3.1

Development of Biosensors in Different Timelines

Year	Development of biosensor
1906	M. Cramer observed electric potential arising between parts of the fluid [20]
1909	Soren Sorensen developed the concept of pH and pH scale [21]
1909	Griffin and Nelson were the first to demonstrate the immobilization of the enzyme invertase on aluminum hydroxide and charcoal [4, 22]
1922	W.S. Hughes discovered a pH measurement electrode [23]
1956	Leland C. Clark, Jr invented the first oxygen electrode [8]
1962	Leland C. Clark, Jr et al. experimentally demonstrated an amperometric enzyme electrode for detecting glucose [9]
1967	Updike and Hicks realized the first functional enzyme electrode based on glucose oxidase immobilized onto an oxygen sensor [10]
1969	Guilbault and Montalvo demonstrated and reported the first potentiometric enzyme electrode-based sensor for detecting urea [11]
1970	Discovery of ion-sensitive field-effect transistor (ISFET) by Bergveld [24]
1973	Guilbault and Lubrano defined glucose and a lactate enzyme sensor based on hydrogen peroxide detection at a platinum electrode [12]
1974	K. Mosbach and B. Danielsson developed an enzyme thermistor [13]
1975	D.W. Lubbers and N. Opitz demonstrated fiber-optic biosensors for carbon dioxide and oxygen detection [14]
1975	First commercial biosensor for glucose detection by YSI [25, 26]
1975	Suzuki et al. first demonstrated a microbe-based immunosensor [27]
1976	Clemens et al. demonstrated the first bedside artificial pancreas [15]
1980	Peterson demonstrated the first fiber-optic pH sensor for in vivo blood gases [28]
1982	Fiber-optic biosensor for glucose detection by Schultz [29]
1983	Liedberg et al. observed surface plasmon resonance (SPR) immunosensors [19]
1983	Roederer and Bastiaans developed the first immunosensor based on piezoelectric detection [30]
1984	First mediated amperometric biosensor: ferrocene used with glucose oxidase for glucose detection [18]
1990	SPR-based biosensor by Pharmacia Biacore [28]
1992	Handheld blood biosensor by i-STAT [28]
1999	Poncharal et al. demonstrated the first nano biosensor [31]
2018	S. Girbi et al. demonstrated a nerve-on-chip-type biosensor for assessment of nerve impulse conduction [32]

density of the DNA strands in the biological sample. The number of DNA strands in a droplet is therefore obtained from [15, 20]:

$$P\left(X_c = k\right) \approx e^{-\lambda}\frac{\lambda^k}{k!}$$

Where k is the quantity of droplet nucleotide bases, and the overall amount of droplet DNA strands is obtained [15, 20].

The process of PCR cannot be adequately performed when a droplet exceeds a modest boundary amount of DNA strands [21]. Droplets containing low DNA strands are called "empty droplets." If after the PCR completion, empty droplets are identified, the time spent operating the heat cycles is squandered. The output signals from the sensors included in the biochip may be used to investigate this issue to decide online if the droplet is empty. In the course of the PCR, as stated in [6, 7], on-chip detectors can assess the strength of photoluminescence in the droplets. If the chance of "the droplet being blank" is high, the droplets will be rejected. A fresh droplet for PCR amplification will then be delivered onto the biochip.

The next part provides prediction algorithms for PCR bioassay to develop an online decision-making system.

3.2.2 IA-Optimized DNA Amplifying Statistical Modeling

The chance of creating a "good droplet" may be calculated during the calibration of the dissipation operation [15]. A good droplet has enough DNA strands to perform the PCR. Let G denote that "the biochip has a nice droplet." A biochip with an unfilled droplet (Gc) is a rare occurrence.

We suppose that the chance of detecting a fluorescence intensity (indicating that the DNA has been replicated) during the ith heat cycle is pi when the PCR operates on a fine droplet. Let Ak denote that "the droplet in the temperature rise does not indicate." The combined chance of G and Ak c is thus

$$P\left(G \cap A_k\right) = \prod_{i=1}^{k}\left(1-p_i\right) \cdot P\left(G\right)$$

If at the kth thermal cycle, no signal is detected, the posterior distribution of "the excellent droplet" (represented as P.G jAk/) will be set as a quotient with G and Ak joint probabilities and Ak likelihood:

$$P\left(G \mid A_k\right) = \frac{P\left(G \cap A_k\right)}{P\left(A_k\right)}$$

$$= \frac{P\left(G \cap A_k\right)}{P\left(G \cap A_k\right) + P\left(G^c \cap A_k\right)}$$

where $P\left(G \cap A_k\right) = \prod_{i=1}^{k}\left(1-p_i\right) \cdot P\left(G\right)$ and $P\left(G^c \cap A_k\right) = P\left(G^c\right) = 1 - P\left(G\right)$.

Therefore, we have:

$$P\left(G \mid A_k\right) = \frac{\prod_{i=1}^{k}\left(1-p_i\right) \cdot P\left(G\right)}{\prod_{i=1}^{k}\left(1-p_i\right) \cdot P\left(G\right) + 1 - P\left(G\right)}$$

$$P\left(G^c \mid A_k\right) = \frac{1 - P\left(G\right)}{\prod_{i=1}^{k}\left(1-p_i\right) \cdot P\left(G\right) + 1 - P\left(G\right)}.$$

3.3 Optimizing Immediately Adjacent Positioning of Resources

We provide our approach to optimize the arrangement under design limitations in this part. The device distance limitations are explained, and the objective function for the performance of components on a PCR biochip is provided. The positioning method for the geometry-based device offers two ways, which may be utilized to choose the optimum device positioning results.

3.3.1 Device-Proximity Constraints on a PCR Biochip

There are three kinds of appliances on the PCR cybernetic implants: reservoirs, transistors for detection, and a thermal unit (e.g., a heater). For avoiding interference between the on-chip devices, it must be presumed that the following restrictions must be met, which impose minimal intrapair (threshold) and interpair separation lengths between the different device pairs. These parameters are supplied by biochip manufacturers as inputs.

3.3.1.1 Separation Constraints

- Restoration tank: The segregation connecting the two tanks should not be smaller than an LRR [22] threshold to prevent fluidic leaking among them. A typical LRR value is four times the circuit length on the DMFB [4].

- DE to DE disconnection: If the DEs are too near together, the immobile droplets may have unwanted intercontamination. The separation from two DEs should thus not be smaller than the maximum LDD value. A standard LDD value is the dielectric quantity of four times the size of the DMFB.

- DE separation reservoir: Fluids put into repositories may have luminous labels, affecting precision in the microscope when evaluating the concentration of the luminous DE [11]. Furthermore, the droplets sprayed from the reservoir may destroy crops to the surfaces of the DE if a DE and a pool are positioned near together. The distance between them should thus not be smaller than an LRD threshold. A standard LRD value is five times the conductor length on the DMFB [18].

- Heater separation tank: As heating can degenerate the tissue material and the reagents placed into tanks [16], no less than a criterion LRH should have been the length between that and storage and the burner. A standard LRH value is five times the electrode length on the DMFB [3].

- DE for separating heaters: The heaters on the cybernetic implants cannot intersect with a DE on the biochip due to the manufacturing process limitation [23]. The distance from the DE to the heater must thus not fall short of the electrode length. The spacing from a DE to a heater must be LDH 1.

For simplicity, we create an operator C, which corresponds to the necessary base station between a pair of gadgets randomly chosen (represented as d_a and d_b):

$$\mathscr{C} : (d_a, d_b) \rightarrow \text{Minimum distance between } d_a \text{ and } d_b.$$

For example, if DE is represented by d_a, and a heater is represented by d_b, then $\mathscr{C}(d_a, d_b) = L_{DH}$. For instance, if the dais DE and the DB heaters are C.da;db/D LDH.

In a PCR biochip configuration for all the components (written as: x1;y1/, X2;y2/;::,.xn ;yn/), we shall define the parameters of ::;..xn;yn/ in order to define the configuration.

$$\sqrt{\left(x_i - x_j\right)^2 + \left(y_i - y_j\right)^2} \geq \mathscr{C}\left(d_i, d_j\right), \forall i, j.$$

3.3.2 Goal Equipment Localization Functionality

- The optimization problem evaluates the effectiveness of device deployment outcomes. In the optimal solution, we examine two metrics:
- The biochip PCR area. The cost of manufacturing the biochip grows. Suppose that all device parameters may be entered as x1;y1/.x2;y2/;::.xn;yn/ for a particular device location G. The biochip area may be calculated as follows:

$$Area = [max\left(x_1, x_2, \ldots x_n\right) - min\left(x_1, x_2, \ldots x_n\right)]$$

$$\times [min\left(y_1, y_2, \ldots y_n\right) - min\left(y_1, y_2, \ldots y_n\right).$$

- The maximum deviation between the heater and the biochip containers. As the heater has a positive temperature coefficient, the specimens in the reservoirs might degrade. If the limitation on the base station between the storage tanks and the heater is not broken, the liquid can be overheated in the storehouses. But it is better to put the reservoir as far as feasible from the heaters to further protect the integrity of the physiological samples in the reservoir. Assuming that all Nr repositories may be marked as xr1; yr1/,xr2;yr2/;::;xrNr; year/ in dimensions for a particular gadget deployment result G, and the heater co-ordinates are Xh;yh/. The defined distance here between the heater and the tanks may therefore be calculated as follows:

$$D_{avg} = \frac{\sum_{i=1}^{N_r} \left[\left(x_{r_i} - x_h\right)^2 + \left(y_{r_i} - y_h\right)^2\right]^{\frac{1}{2}}}{N_r}.$$

The mathematical formulation for biochip design optimization (F.G/) may be calculated as follows for the given equipment placing outcome G:

$$F(G) = \alpha Area + \frac{\beta}{D_{avg}(G)}$$

Where alignment and alternate weighting are user-determined characteristics that may be modified to take into account the comparative value of each measure. If F.G1/<F.G2/, G1 shall be seen as positive performance in two device deployment outcomes, G1 shall be regarded for G1 and G2.

3.3.3 Positioning of Equipment on a Biochip PCR

There are two features for the device positioning challenge in a biochip:

- Commonly, the design of an e-fluidic digital biochip design and a 2D range with the same electrodes. We may thus assume that the electronics and electrodes remain at integer data points alone.

As each gadget pair might have a different limitation on the separating pacing of each device, the entire computational complexity may be vast, and we assume that it's an NP-hard issue to discover a minimal area layout [24]. This part proposes a greedy approach based on an iterative process for developing our design tool.

The PCR pattern should be created on a "digital plane" with all coordinated integer points. The digital planes can be numerically expressed as Z2, where Z is an arithmetic setting. A "digital point" in Z2 is believed to be positioned in the middle of each portable computer.

When DA and DB are put on the chip for connected devices, we draw a circle with C.da; DB/ radius centering da. The area inside the ring is described da as the "forbidden area" of DB. All electrodes, which overlap permanently or temporarily with the prohibited area of DB, are denoted as O.da; DB. Note that in a collection of visual points, the O.da; DB/ may be seen as the "digital object" in Z2 [25]. The digital item that matches the heating is indicated.

In other words, when the reservoir is located on the body or outside the physical product of the heater, the appropriate separation requirement is fulfilled for the pools and the heating element. All components must be positioned as near as possible to reduce the chip area. An example of a correct and compacted positioning of R2 in the vicinity of H, DE D1, and R1. This is a digital object with the isostatic convex envelope fiction of three prohibited areas (H, D1, and R1, respectively, regarding R2). Please note that the R2 tank should not be placed in this item for the proper context; it should be positioned as carefully as possible for its compaction to reduce the whole surface area of the chip, including all of the components. All PCR biochip equipment are included in the ready queue, Qdev, during the first step of device installation. Device stacking is random in Qdev.

This technique for placing a device is shown in Figure 5.3b, where a new R2 reservoir may be appropriately positioned in the context of an H, DE D1, and an R1 pool. We also suppose that the heater is the first step in our algorithms at the beginning of the construction.

We selected the gadget from the head of the resource provisioning Dev. for every successive stage. In all previous versions, the gadget has been well positioned in terms of its capabilities.

3.3.4 Machine Positioning Parameter Optimization

- The results of placing in a particular sequence may be determined from the placer of the devices that fulfill all the limitations on the closeness of the instrument. Suppose we have no distinctive reservoirs, no distinctive DEs, and Nh heaters. The amount of feasible job scheduling including all equipment (i.e., the number of decision variables for PCR biological chip deployment) is therefore shown by:

$$\frac{\left(N_r + N_d + N_h\right)!}{N_r! N_d! N_h!}.$$

- We may use two different techniques to pick the "best" option based on the provided objective functions.

3.3.4.1 Listing All Conceivable Gadget Arrangements

- As a result of its high manufacturing costs, just a few EDs are incorporated into a biochip, and only a single refrigerating module is enough. In a standard mixing process following a reverse transcription, the number of compounds required is around eight. The output outlets of the capacitors are customarily placed around the chips and the heaters within. Therefore, a management system considerably decreases the number of viable orders. We may thus carry out all the orderings of the previously stated heuristic method. For each gadget, the value of the inserted optimization problem may be computed. An "optimal solution" is identified for placing corresponding to the minimal value of the optimization problem.

3.3.4.2 Deployment of Computer Simulation Ring-Based Devices

The rising number of devices on a PCR biochip might make computing too expensive to be effective. In this situation, we offer a technique based on genetic algorithms (GA) to find an ideal position of the instruments.

Each initial study includes all PCR-based devices and is encoded in the GA-based method as a genetic material. The following defined random keys have chromosome vectors (or "genes"):

$$Chromosome = (gene(1), gene(2), gene(3), ..., gene(n))$$

The sampling number from Œ0;1 is where n is the instrument size to be adjusted and the genes i/ I've been 1 to n). The device i's precedence value is set to Pdev.i/D gene.i/.

The focus on process thinking lists all devices generated by sorting the priority values. The placer of the device then introduces a work placement that can be created using any chromosome to meet the nearness restrictions.

4n chromosomes are produced haphazardly at the commencement of the GA-based method. The heuristic method takes into account the following techniques called:

- Reproduction. The chromosome that matches the lowest cost value is transmitted to the following generation.
- Crossover. Two-parent chromosomes are selected randomly from the aging population during the crossover operation. In the following manner, the offspring chromosome is produced. The gene of the chromosome's offspring can be either replicated from the parent chromosomal gene. The likelihood of gene inheritance i/ in the father and mother is determined by P and 1 P correspondingly. The chromosomes of the future generation's kids are being examined by the civilization breeding them.
- Mutation. The population's new chromosome is haphazardly produced to ensure population variety.

The set of chromosomes in each generation remains 4n in the GA-based heuristic. In the evolutionary process, the n best chromosomes are replicated into the next generation, the chromosome equivalent to the most minor and cost density function. In the initial demographics, the original population developed 2n chromosome, and unpredictable media generated an additional n genetic material. This % can be further enhanced by testing. After multiple decades of development, the lowest-cost chromosome is selected from the final subpopulation. The solution for positioning the gadgets is built using the chromosomal set. In the framework of a specific PCR procedure, we discuss how to improve droplets' transit and mingling.

3.4 Deployment of PCR Biosensor Immunoassay Reservoirs

Several healing potions must be combined in a specified ratio following a standard mixing method after DNA amplification. The number of mixed divides is dependent on the algorithms and the objective ratio. Combining operation is typically described in a blending diagram G., two examples of the infrastructure fund graph [26]. The bioassay sequence diagram is guided. Every null in-grade node in the network indicates a dissipation operation and the fluidic processes (e.g., mixing and diluting) is the interior node [27]. The travel costs for droplets in the PCR should be selected depending on the supplied bioassay sequenced architecture which needs the proper relocation of reservoirs and combining/dilution planning as outlined in the sequence graph.

The conceptual design technique for the PCR brain chip will be discussed in this part. The low-cost biochip "electrode ring" construction is implemented. A low-cost PCR biochip droplet transportation problem is presented. It contains the bioassay algorithm for the assignment of PCR biochip resources.

3.4.1 Low-Cost Biosensor Electrolytic Rings

For each of the gadgets on a 2D grid, the positioning method described above allocates a position. The heater generally places reservoirs at the middle of the arrangement and is located on the edge of the biological chips. The design process of the PCR biochip has to be completed by determining the locations of the electrodes on the biochips and establishing droplets between the devices, as shown in Figure 3.1.

The output terminals of all reservoirs are interconnected by a "cycle" of electrodes to decrease conductors in the usual low-cost bio channel [7, 8]. The mixer modules are not predestined to have a particular zone; every combining action occurs on the electrode between the reservoir output ports. Let D1 and D2 indicate two distributing procedures of reservoirs R1 and R2. Let M1 symbolize a mixing process between those two droplets, where droplets of R1 and R2 are combined in the antenna structure between these two electrodes.

The load balancing of PCR biochips can be increased by positioning the reservoir output on the circumferential electrode route. The arrow shows their original transport

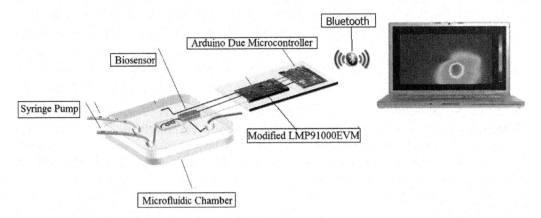

FIGURE 3.1
Working of PCR biosensor.

pathways, and the two drops must be combined. The travel trajectories of these two droplets are transformed when a fault exists on the pattern. Therefore, the PCR biochip may perform bioassays with an alternate droplet travel path, notwithstanding the defect.

3.4.2 Low-Cost PCR Biosensor Particle Sequencing

All reservoirs are regarded as undifferentiated gadgets while assessing the location of equipment. We must nonetheless assign reservoirs to every specimen/reagent employed in a bioassay cytotoxic activity on the PCR biosensor to be performed.

If a cheap biometric scanner is available, an "electrical ring" shape must do numerous activities simultaneously; the drip routing pathways may clash. Here we suppose that x1 to x7 reagents are kept in tanks R1 to R2.

According to the sequence graphs, the drips discharged from the R1 and R4 storage tanks must be mixed in operation M5, and the raindrops released from R3 and R5 must be combined in the M3 system. We suppose that output locations are arranged in the same way in R1, R3, R4, and R5. The particles dispensing from R1, R3, R4, and R5 must be carried in the order specified by the arrow to execute mixing operations M3 and M5. It may be noted that these four droplets cannot be taken simultaneously on the biochip to avoid head-on accidents. Therefore, M3 and M5 mixing can't be performed simultaneously.

By altering the reservoir assignment settings, droplet transportation conflicts can be eliminated. Assume that the R1 R7 reservoir outputs are positioned, the degree of parallelism will rise for the procedure. The subtree methods with the M9 root node (as described as SubtreeM9) include just the droplets of the R1 and R4 tanks. Thus, all SubtreeM9 operations may be carried out on transistors immediately linked to the R1 and R4 output ports. Since there is no additional reservoir output port in this location, droplets with R1 and R4 dispenses can be combined without droplet routing problems.

Similarly, activities can be conducted in the areas listing electrodes connecting the analog inputs of the reservoirs in subtreeM8, subtreeM13, and subtreeM14. Ri, Rk is spelled as Regions. Ri is represented as Continent. There is no conflict between droplet routings, and hence, all blending activities in the mixture tree may be carried out locally.

3.4.3 Deployment of Virtual Screening Reservoirs

The findings the consumer makes an assignment of the reservoirs, droplet routing, and planning for fluid management may be determined as follows based on this discourse.

Sequence diagram cluster formation: If the substring SubG operation includes droplets from just two reservoirs, all the SubG activities between these two liquids can then be conducted in the region. At this point, we thus need to determine the maximum subsection on the sequence chart, so there are only two ingredients of RSi and RSJ for operations within the subsection. In the sequenced graph, this number of pixels is described as "basic mingling clusters." We monitor the significance of the resulting droplet "components" based on the nodes representing the bioassay's final blending procedures (i.e., null-out-degree nodes). For instance, the M18 output droplet combines four R1, R2, R3, and R4 reactants. Then we have to retrace the previous M18 processes and then inspect their resulting droplet for components. Until the N operation is found, this trace-back process repeats. There are just two significant elements in their respective weight's droplets. A foundational sequential mixing group is the subparagraph comprising operation N and its antecedent.

Allocation of the reservoir: The bioassay-specific assignment of the pool may be established after discovering all bare mixture clusters of the sequenced diagram. For every cluster mixing. The RSi; RSJ/specimens will be laid out as "neighbors" for the RSi and RSJ respective reservoirs. An explanation of the neighborhood concept is provided as follows:

> Definition 1. Suppose a set of pools is present, SR D fRn1, Rn2, Rn3, Rnk g, and they are all arranged on a circle of electrodes. Pairs Rni and Rnj (n1 ni < nj nk), where the electrode route P.Rni; Rnj / travels through the Ri and Rj output ports, have "relatives" in the SR, and on this cathode route, there is no output port of other SR reservoir configured.

> Definition 2. Rnj will be the "neighbor counterclockwise" of Rni if you move from Rni to Rnj along the P.rni; Rnj/electrode route in a clockwise manner. Rnj is the "neighbor" against the clock in Rni, otherwise.

The following is an example. The histogram may be broken down into five major groups of mixing. An indifference curve can indicate the link between these mixing groups. A reservoir symbolizes each node in the network; when the same cluster has two pools, the edge between node pairs matching those of storage tanks is provided. In the related unspoken graph, each gout contains two alternative routing instructions, like low-cost biochips' electrode ring. If there are two fundamental mixing groupings in a reservoir, droplets that have been discharged from the storage tank cannot be employed simultaneously in all these clusters during mixing. Therefore, the undirected graph removes a portion of the connections such that every node has no more than two parameters. The following guess is as follows: Conjecture 5.1. Suppose that GC reflects the mixing ratio of the clusters in a bioassay, but partial derivatives are in each node in GC. Then we can always position the reservoir on the ring of the electrodes so that every pair of pools among them may be placed next door.

When there is a rim between them, we position Rni as the neighbor of Rnj for two reservoirs Rni and Rnj. As mentioned above, there are two neighbors for each reservoir on the electrode rings: clockwise, outwards, and anti-clockwise. If another pool is in Rnj's nearby places (e.g., clockwise neighbor), we place Rni in an empty location. Notice that each edge is not beyond two borders; therefore, no additional reservoirs occupy both neighboring areas of Rnj.

Operational planning: The sequencing of activities may be established based on the sequenced network once the storage allocation phase has been finished. As activities may take place locally without long-distance transfer in each core mixing cluster, transactions can be done simultaneously in several groups. The planning of activities within the dependency among outputs and inputs might be predicated on the same clustering [28].

At the beginning of the bioassay, procedures included with the fundamental blending groups are conducted. After the blending cluster activities are done, more operations are performed. For example, M1 and M3 are in the entire cluster's mixture in Figure 5.4 R1; R4/. After cluster procedure. R1; R4/ when done, one of the clusters. R1; R4/output droplets are mixed with M8-Droplet R2, cluster output droplets. Cluster R1; R4/. In operation, M9, R2, R4/ will be combined. The programming of all the hydraulic steering action may be obtained from this technique in the sequencing graph.

3.5 Particle Localization Accessibility

The surveillance of particles is crucial in a computerized maintenance management digital microfluidic system. Investigators typically employ fluorescence cameras and imaging technology placed on the computing platforms to observe many activities concurrently on the biochip [29, 30]. The pictures obtained may be used to monitor the quantities and percentages of the tiny particles at each phase of the experiment and accurately estimate the time necessary to accomplish vacuum distillation procedures on a biochip [29, 30].

The abuse of the top-view picture will nevertheless contribute to incorrect findings for the final adjustment amount. For instance, for 10s, a fluorescence droplet is fused with a non-fluorescein droplet, and the fluorescence microscopy captures combined droplet pictures from top and side views [30], respectively. The fluorescence reagent can be seen in the combination droplet uniformly distributed from the top, i.e., the mixing process might look completed. The merged droplet has several layers from the side profile, which is insufficient for the mixing procedure. Cyber-physical microfluidic biochips from many viewpoints are so essential to monitor. On the computing platform, digital sensors may be fixed so that pictures can be obtained from two distinct views. If many droplets are handled simultaneously on the biochip, a side-view evaluation should consider the droplet's transparency. It is assumed droplets 14 are combined in four biochip mixers concurrently, with arrows specifying the vehicle's motion. We suppose that these four droplet-combining procedures start simultaneously with the droplets in the same orientation.

Droplets 1 and 3 are positioned in one line of the range of the electrode, and in the same line of the electrode, there are droplets 2 and 4 in the starting position. When the 1 and 3 movements are planned onto the y-axis, they always intersect during the mixing process. Therefore, when the cameras are positioned, and the side views of all four showers are captured, droplet 1 is always hidden under droplet 3. The monitoring system will see droplet 1 as "invisible." Drop 2 is concealed below droplet 4, and the monitoring mechanism is likewise invisible. Likewise, when the camera is fixed and the four drips are monitored on the side of the camera, drops 2 and 4 are always invisible to the monitoring program. To detect all the combining operations concurrently, the condition that "some of the raindrops are transparent for the camera or fluorescence microscope" must be prevented. We may specify multiple starting locations for droplets to address this drawback. The connection between time and the position of these two raindrops is projected on the x-axis by the mobility of droplets 1 and 2. Droplets 1 and 2 do not overlap at any point; thus, the camera may take pictures of both of these goblets. The most significant number of droplets visible from the side view may be further determined by projecting the motions of droplets on the x- or y-axis.

For example, we may calculate the maximum number of raindrops that can be seen in parallel for 14 mixers that overlap the projection on the y-axis by selecting different beginning locations and traveling orientations.

In designing fluid treatment operations on a microfabrication biosensor, this transparency restriction of nanoparticles should be considered. The simulation program aims to check the accessibility of each droplet for the movements of the droplets toward this x- or y-axis. If a transparent droplet is available, the biosensor detection systems modify the original positions of the droplets.

3.6 Cyber-Physical and Layout-Aware PCR Biochip

This subsection shows the actual simulated results to validate the technique proposed for developing a cyber-physical and layout-conscious PCR biosensor. Then effects of the tests are submitted for three protocol blending benchmarks.

3.6.1 Polymerase Chain Probability Monitoring

The proposed mathematical formulation may be used for online decision-making during the DNA nanoparticle amplification. It is essential to highlight that those fluorescent indications from multiplex PCR cannot be instantly detected when the targeting DNA strands are duplicated. This is due to each recognition method's anticipated noise signal (MDS) [28]. The plasmid fragments' fluorescence signal is not visible if its amplitude is less than the MDS. We thus assume that the probability of the fluorescence intensity on extended DNA strands before the Nth combustion stroke is zero. The odds of estimating the fluorescence channel (i.e., the DNA is enhanced) are highest.

If we suppose that the connection between the amount of gamma irradiation carried forward) and P. GcjAi/(i.e., the likelihood of that "droplet being vacant").

The equivalent P.GcjAi/ connection in this case. The value of p is 104, 105, and 106 in the three curves, correspondingly. Figure 10.5 shows the likelihood that if "this droplet is an unfilled droplet," it is set to 105 if no signal at the 36th melting process exists, 60%; if there is no signal at the 37th melting process, it is likely to become 99%. Thus, if no sign is found after heat cycle 37, we have a confidence of 99% that the droplet is sufficient and should be eliminated for the PCR DNA strands.

A complete prediction model may be used to assess the reverse transcription stage. As in the ideal case, the number of stranded DNA in a droplet increases exponentially; the PIM value is high.

$$p_i = \begin{cases} 0, & \text{if } i < N \\ p* \times 2^{i-N}, & \text{if } N \leq i \leq N - \log_2 p* \\ 1, & \text{if } N - \log_2 p* < i \end{cases}$$

3.6.2 PCR Bioelectronics Design and Structure

One of these designs is minimal in the form of "R," "D," and "H," correspondingly, representing the analog inputs of the tanks, the PDS, and the heater. If the unit area of each electrode is considered to be the shortest rectangular surrounding, the layout rotates.

After designing a minimal area architecture, a specific PCR technique is established by allocating an appropriate reservoir to decrease droplet movement. A solution mixing method is referred to as Bioassay 1 [26]. The genetic algorithms may obtain the best outcome of the assignment of the reservoir according to the regularity of use in the mixing operation. The final architecture achieved after selecting the pool reduces the travel time of droplets. R17 is equipped with x17 reactant. During the bioassay, mixer paths can be allocated to the electrode connecting the output ports of a pair of reservoirs. Therefore, all mixing procedures in the protocol may be done "locally" without problems in routing droplets via vast distances. For example, M1, M4, and M8 are reagent x4 and x7 inputs, which are to be conducted on electrodes somewhere between

R4 and R7 correspondingly. The degree of concurrency in the fluid treatment procedure is also high because conflicts may be avoided between the transfer of ideas and droplet routing in the biochip. Here we suppose the mixing duration on a 1 4 mixer is 5s. And the traveling time of the droplet from electrodes is tms (tm usually varies from 0.01 to 1 [31]). Then it's computed in (35+36tm) seconds that the entire biomass essay must be performed.

Here we develop a further positioning of the gadget as the basic algorithm. The primary approach is to arrange all devices at the layout of one boundary (including exit ports for reservoirs, PDs, and heats), LMAX D'MAX {LRR, LDD, LRD, LRH, LDH} The distance between two devices is specified. All these devices are connected via a circular route, consisting of all the border electrodes.

The baseline technique size of the antenna structure is 20 15, whereas the necessary electrode number is 68. The whole bioassay is carried out with .80 C 115tm/s.

Thus, the suggested design technique may lower the size of the chip by 59.7% and 35.3%, and the executable time by 57.8%, compared to the previous algorithm (when the value of tm is set as 0:1).

We then investigate a technique representing a genuine PCR mixing ratio [32]. This biological test is known as physical test 2. In the eight components, the mixing balance is written as follows:

In addition to this, Bioassay 3 is a mixing technique with three sample preparation. We are considering this. Bioassays are enhanced considerably in layout size, electrode count, and prediction accuracy compared to the baseline procedure in all three situations.

Since the number of devices to be put is very modest, these three laboratory bioassays maximize the experimental findings of the method suggested. We are also developing three benchmarks with comparatively many algorithms, and the fundamental algorithms are used to obtain the results. The simulated results are provided. It should be noted that the approach, due to the high computational complexity, is inaccessible while finding the optimal outcomes of device placement for benchmark tests 2 and 3. The approach sets the constants èf and èf correspondingly as 1 and 10. Compared with the baseline method, we find that approach 2 can lowers the biochip area by 64.1–72.9%.

3.6.3 PCR Biochip Designing Deficiency Compatibility

Physical flaws may arise on DMFBs due to manufacturing errors and electrode deterioration. These faults can be categorized according to their placement on the biochip in two groups. Suppose a defect disengages the droplet path to two "isolated portions" or the deficiency intersects on the biochip with a biochip storage tank. In that case, the error is classified as a "catastrophic defect" (potentially fatal defect). All other deficiencies are categorized as "non-disaster flaws."

The PCR biochip cannot be utilized further in the event of a disastrous failure. The most vulnerable parts of semiconductors in terms of pathophysiology experimental method are found. The total number of terminals where defects are devastating for each of the three PCR biosensors constructed for bioassays for each of the three PCR biosensors constructed for bioassays.

If there is a failure to resolve a calamity, we can forward the droplets to their destination. The re-routing of the droplets can affect the road strength and reduce the population administration parallelism. It should be noticed, therefore, the execution period of the immunoassay on a biochip with a non-disaster flaw is more time-consistent with the defect-free biochip.

For each PCR biochip, a non-catastrophic flaw is randomly inserted into the biochip, and the measurement duration is then calculated. The defective insertion simulation has been carried out for all potential non-catastrophic faults in the layout—the accurate temperature and confidence interval for the performance of bioassays on defective biochips. The tm value is set to 0.1. The proportion of electrodes with disastrous faults ranges from 33.3% to 52.3% from the table. The PCR biochip may be employed by graceful degradation with 27.267% increase in the execution duration of bioassay with just the existence of a single non-catastrophic flaw.

3.7 Conclusion

These chapters have shown that fundamental in liquid operations, such as supply and temperature cycling, notwithstanding the uncertainties, cyber-physical connectivity in digital microfluidics may be employed for reliable bioassay on the chip. In response to data provided by sensors, the suggested design method allows dynamic, informed decisions online for the ending of heat cycles. We have also launched new device arrangements and arrangement designs to protect the array from undesired device noise and interference, minimize conflicts in droplets' routing, and decrease bioassay runtime. Droplet transparency is also addressed during the planning of liquid activities in the monitoring program. The simulated data from lab procedures show that the suggested design approach may obtain significant dependability and fault tolerance with reduced chip sizes and electrode counts.

References

1. Y. Luo, B. Bhattacharya, T.-Y. Ho, and K. Chakrabarty. "Optimization of a polymerase chain reaction on a cyber physical-digital microfluidic biochip". In *Proceedings IEEE/ACM International Conference on Computer-Aided Design*. pp. 622–629. (2013).
2. J. Lage, J. Leamon, T. Pejovic, S. Hamann, M. Lacey, D. Dillon, R. Segraves, B. Vossbrinck, A. González, D. Pinkel, D.G. Albertson, J. Costa, and P.M. Lizardi. "Whole-genome analysis of genetic alterations in small DNA samples using hyperbranched strand displacement amplification and array-GH". *Genome Research*. Vol. 13(2). pp. 294–307. (2003).
3. J. Berthier. *Micro-Drops and Digital Microfluidics*. Norwich, NY: William Andrew. Vol. 6. pp. 633–640. (2008).
4. I. Erill, S. Campoy, J. Rus, L. Fonseca, A. Ivorra, Z. Navarro, J. Plaza, J. Aguilo, and J. Barbe. "Development of a CMOS-compatible PCR chip: Comparison of design and system strategies". *Journal of Micromechanics and Microengineering*. Vol. 14(11). pp. 1–10. (2004).
5. C. Zhang, and D. Xing. "Miniaturized PCR chips for nucleic acid amplification and analysis: Latest advances and future trends". *Nucleic Acids Research*. Vol. 35(13). pp. 4223–4237. (2007).
6. D. Brassard, L. Malic, C. Miville-Godin, F. Normandin, and T. Veres. "Advanced EWOD-based digital microfluidic system for multiplexed analysis of biomolecular interactions". *IEEE International Conference on Micro Electro Mechanical Systems (MEMS)*. Vol. 3. pp. 153–156. (2011).
7. D. Jary, A. Chollat-Namy, Y. Fouillet, J. Boutet, C. Chabrol, G. Castellan, D. Gasparutto, and C. Peponnet. "DNA repair enzyme analysis on EWOD fluidic microprocessor". *Proceedings of the NSTI-Nanotechnologies Conference*. Vol. 2. pp. 554–555. (2006).

8. K. Chakrabarty, and F. Su. *Digital Microfluidic Biochips: Synthesis, Testing, and Reconfiguration Techniques*. Boca Raton, FL: CRC Press. (2006).
9. Y. Luo, K. Chakrabarty, and T.-Y. Ho. "Dictionary-based error recovery in cyber-physical digital-microfluidic biochips". In *Proceedings IEEE/ACM International Conference on Computer-Aided Design*. pp. 369–376. (2012).
10. R. Liu, J. Yang, R. Lenigk, J. Bonanno, and P. Grodzinski. "Self-contained, fully integrated biochip for sample preparation, polymerase chain reaction amplification, and DNA microarray detection". *Analytical Chemistry*. Vol. 76. pp. 1824–1831. (2004).
11. L. Malic, T. Veres, and M. Tabrizian. "Detection of DNA hybridisation on a configurable digital microfluidic biochip using SPR imaging". *International Conference on Miniaturized Systems for Chemistry and Life Sciences*. Vol. 5. pp. 829–831. (2008).
12. L. Malic, T. Veres, and M. Tabrizian. "Two-dimensional droplet-based surface plasmon resonance imaging using electrowetting-on-dielectric microfluidics". *Lab on a Chip*. Vol. 9. pp. 473–475. (2009).
13. J.S. Dhatterwal, K.S. Kaswan, and Preety. "Intelligent agent-based case base reasoning systems build knowledge representation in Covid-19 analysis of recovery infectious patients" in a book entitled "application of AI in COVID 19" accepted in Springer series: Medical virology: From pathogenesis to disease control. (2020). (e-Book), 978-981-15-7316-3 (Hard Book). https://doi.org/10.1007/978-981-15-7317-0.
14. S. Koster, F. Angile, H. Duan, J. Agresti, A. Wintner, C. Schmitz, A. Rowat, C. Merten, D.Pisignano, A. Griffiths, and D. Weitz. "Drop-based microfluidic devices for encapsulation of single cells". *Lab on a Chip*. Vol. 8(7). pp. 1110–1115. (2008).
15. R. Daniel, M. Dines, and H. Petach. "The denaturation and degradation of stable enzymes at high temperatures". *Biochemical Journal*. Vol. 317(1). pp. 1–11. (1996).
16. F. Ji, M. Juntunen, and I. Hietanen. "Evaluation of electrical crosstalk in high-density photodiode arrays for X-ray imaging applications". *Nuclear Instruments and Methods in Physics Research*. Vol. 610(1). pp. 28–30. (2009).
17. R. Evans. "Optical detection heterogeneously integrated with a coplanar digital microfluidic lab-on-a-chip platform". *Proceedings IEEE Sensors Conference*. Vol. 9. pp. 423–426.
18. S. Koester, L. Schares, C. Schow, G. Dehlinger, and R. John. "Temperature-dependent analysis of Ge-on-SOI photodetectors and receivers". In *IEEE International Conference on Group IV Photonics*. Vol. 6. pp. 179–181. (2007). (2006).
19. C. Zhang, and D. Xing. "Single-molecule DNA amplification and analysis using microfluidics". *Chemical Reviews*. Vol. 110(8). pp. 4910–4947. (2010).
20. D. Woide, A. Zink, and S. Thalhammer. "Technical note: PCR analysis of minimum target amount of ancient DNA". *American Journal of Physical Anthropology*. Vol. 142(2). pp. 321–327. (2010).
21. J. Webster, M. Burns, D. Burke, and C. Mastrangelo. "Monolithic capillary electrophoresis device with integrated fluorescence detector". *Analytical Chemistry*. Vol. 73(7). pp. 1622–1626. (2001).
22. U.-C. Yi, and C.-J. Kim. "Soft printing of droplets pre-metered by electrowetting". *Sensors and Actuators Part A: Physical*. Vol. 114(2–3). pp. 347–354. (2004).
23. M. Garey, and D. Johnson. *Computers and Intractability: A Guide to the Theory of NP-Completeness*. W.H. Freeman & Company. Vol. 12. pp. 45–60. (1979).
24. P. Bhowmick, A. Biswas, and B. Bhattacharya. "ICE: The convex aesthetic envelope of a digital object". In *International Conference on Computing: Theory and Applications*. pp. 219–223. (2007).
25. S. Roy, B. Bhattacharya, P. Chakrabarti, and K. Chakrabarty. "Layout-aware solution preparation for biochemical analysis on a digital microfluidic biochip". In *Proceedings IEEE International Conference on VLSI Design*. pp. 171–176. (2011).
26. Y. Luo, K. Chakrabarty, and T.-Y. Ho. "Design of cyber-physical digital-microfluidic biochips under completion-time uncertainties in fluidic operations". In *Roc. IEEE/ACM Design Automation Conference*. pp. 44–50. (2013).
27. Y.-L. Hsieh, T.-Y. Ho, and K. Chakrabarty. "Design methodology for sample preparation on digital microfluidic biochips". In *IEEE International Conference on Computer Design*. pp. 189–194. (2012).

28. P. Paik, V. Pamula, and R. Fair. "Rapid droplet mixers for digital microfluidic systems". *Lab on a Chip*. Vol. 3(4). pp. 253–259. (2003).

29. P. Paik, V. Pamula, M. Pollack, and R. Fair. "Electrowetting-based droplet mixers for microfluidic systems". *Lab on a Chip*. Vol. 3(1). pp. 28–33. (2003).

30. E. Bolton, G. Sayler, D. Nivens, J. Rochelle, S. Ripp, and M. Simpson. "Integrated CMOS photo-detectors and signal processing for very low-level chemical sensing with the bioluminescent bioreporter integrated circuit". *Sensors and Actuators Part B, Chemical*. Vol. 85(1–2). pp. 179–185. (2002).

31. Preety, K.S. Kaswan, and J.S. Dhatterwal. "Securing big data using big data mining". In Data *Driven* Decision *Making* Using *Analytics*. UK: Taylor Francis, CRC Press, 2021.

32. Y.-L. Hsieh, T.-Y. Ho, and K. Chakrabarty. "A reagent-saving mixing algorithm for preparing multiple-target biochemical samples using digital microfluidics". *IEEE Transactions on Computer-Aided Design of Integrated Circuits and Systems*. Vol. 31. pp. 1656–1669. (2012).

4

Security and Privacy Aspects of Cyber Physical Systems

Chhavi Sharma

CONTENTS

4.1 Introduction

Cyber physical systems (CPS-type receptor) play an important role in the sector of the Internet of Things (IoT) and Industry-v4.0. The CPS allow intelligent application and services, in order to properly function, in real time. The integration of cyber physical systems for the exchange of different types of data and the confidentiality of the information in real time [1]. The CPS receiver is a network of in-house systems. This is a network of internal systems, which interact with the physical inputs and outputs. The CPS-type receiver has three main components: sensors, collectors, and actuators. The CPS is the backbone of

DOI: 10.1201/9781003220664-4

TABLE 4.1

CPS Description and Classification

Naming	Classification	Description
Smart house	Industrial consumer IoT	Control smart devices
Smart grid	Industrial IoT	Smart efficient energy
Oil refinery	Industrial transportation IoT	Naphtha, gasoline, diesel
Water treatment	Industrial consumer IoT	Improved water quality
Medical devices	Medical wearable IoT	Enhanced medical treatment
Smart cars	Industrial transportation IoT	Enhanced driver experience
Supply chains	Industrial transportation IoT	Real-time delivery

the patient's ability to respond to new conditions. The CPS-type receivers, the system can bring the environment to adapt to and monitor the physical world [2–, 13] (Table 4.1).

The CPS-type receiver systems are widely used in many fields [photo 1]. The CPS-type receiver systems, communication systems, for example, the power transmission systems, communication systems, agricultural/environmental management systems, military systems [7, 8], and the autonomic system (the system for unmanned aerial vehicles, robotics, autonomous vehicles, etc.) [9, 10]. The CPS-type receiver can be used in the health care system in order to improve the quality of health care [11], and a supply chain management system is to provide an eco-friendly, reversible, cost-efficient, and safe production process.

The CPS receiver system is heterogeneous in nature. The CPS-receiver systems, the confidence of a personal and confidential information. The CPS-type receiver system has a wide range of locations. The system is subjected to physical security threats, attacks, and challenges.

4.2 Characteristics of CPS

The recipient of the CPS type consist of in-house systems, real-time systems, networks, and control theory.

4.2.1 Embedded Systems

The internal systems of several computers to interact directly with the physical world (sensors, actuators, or actuators) need the CPS-type receiver system with limited resources. They don't have the total computational power of classical computers. In some of our internal systems, while work experience is not running, the operating system, but it is just that you are working inside the program. The firmware may not be modified or deleted by the end user without the need to use special programs and remains in the unit, irrespective of whether it is on or off.

Cyber physical systems, which allow for the monitoring and control of complex physical processes with the help of a computer and the IoT technology are becoming increasingly relevant in the major application areas, including manufacturing automation and smart cities.

4.2.2 Real-Time Systems

Real-time systems is that the system is real-time, which means that the system is subjected to real time, that is, the answer is, they can't guarantee that, within a specified time, the system needs to comply with the time limit referred to in it. For example, flight control system, real-time displays, etc.

4.2.3 Network Protocol

A formal language is compulsory for the communication between the transmitter. For speech, in order to be effective, it is important to agree on a common language between the two of communicating with people. This means that language is an important part that will give the necessary rules of social communication.

The task is to turn on the unit, the unit of communication, that is, the general provisions laid down in the series and to the feedback. This is where some of the protocols come into play. It defines a set of rules that are required for the communication to work as expected (Figure 4.1).

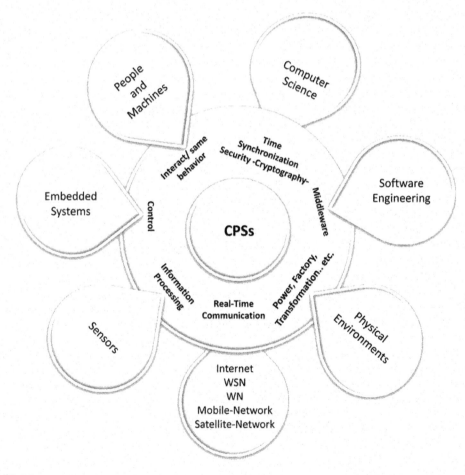

FIGURE 4.1
CPSs are used in multiple domains.

4.3 General Architecture of Cyber Physical Systems

Figure 4.2

4.4 Defining Security and Privacy

Security is one of the set of measures that are designed to ensure that the system is able to reach the intended target, but not the negative behavior. New features may be added to the system in order to enhance the functionality in order to enhance the functionality [18–32].

When a function is connected to the system, it must be secured in such a way that it does not compromise on the intended function(s) and in the face of new attacks. The concept of security and privacy is for appropriate use and protection of your information. Integrity is the ability to limit your ability to express yourself. In general, it is often considered to be a security issue. Privacy is a difficult part of your life. An operating system must protect the privacy of its users. Ensure confidentiality and security, which cannot be communicated to unauthorized persons, but also of privacy, as it allows the owners of their own, in order to regulate the dissemination of information.

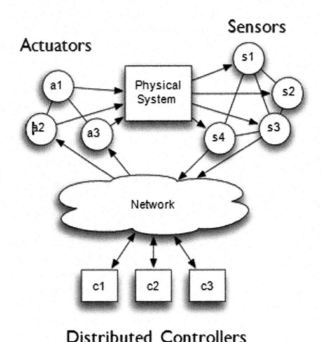

FIGURE 4.2
Architecture of cyber physical systems [12].

4.4.1 Cyber Security and Privacy

The CPSs comprise the information technology and physical dimensions. The concepts of data privacy and security can be applied to both and the individual CPSs. That is, the central committee of the Central Public Sector Undertaking (CPSU) and some of the security measures to be applied. Cyber security is a combination of attack, prevention, detection, suppression, recovery, and clean-up in order to accommodate a risk-management strategy.

The three core principles of information security, is "the CIA Triad", confidentiality, integrity and availability. Integrity, is only authorized users can access resources related to CPS.

Disabled-the Assets are accessible to authorized parties at appropriate times.

The CIA triad provides secure access to the right information to the right people. Principles of data privacy:

Identification: The process or action of an identifier, or a user process.

Non-repudiation: Non-repudiation is a presentation of irrefutable evidence that a message may be sent or received. If a message or a transaction can be challenged, in a major event identification, which can be appealed, and the benefits.

4.4.2 Physical Security and Privacy

The following principles are designed to provide shelter:

Deterrence: Deterrence is the act of preventing an act or event that, by suggestion, doubt or fear the consequences.

Detection: Finding a positive rating to a particular item that can cause a potentially harmful action.

Delay: The delay time on the obstacles slows down or prevents the enemies, you will be limited in your options and may be compelled to take an adverse action.

Response: The answer is action—action taken by the appropriate forces on the ground, and for a period of time in order to stop the enemy advance.

Neutralization: The neutralization of the armed forces of the Russian Federation, to deprive the enemy of the ability to take action or any other activity.

4.5 CPS Vulnerabilities, Threats, Attacks, and Failure

There is a broad range of IoT devices that can use the receiver's CPS. IoT domains are interacting using a variety of techniques and protocols. In the CPS-type receiver, the system is vulnerable to threats and attacks.

4.5.1 CPS Security Threats

Threats to the security of CPS receivers can be classified as cyber-physical threats:

4.5.1.1 Cyber Threats

The information technology threats: The main focus is in the area of industrial security on the internet was to focus on cyber threats, rather than a natural hazard.

Cyber security can't be assessed; however, we can analyze it from different points of view, as shown in Tables 4.2 and Table 4.3.

Some of the threats in the context of CPS, the type of recipients include:

1. **Function**: It consists of concealing the identity of a reliable, is actually damaging an unknown source. In this case, the attack is able to replace the sensor, for example, by sending you on a bad, misleading, and/or inaccurate data in measuring, control of the center.

2. **Sabotage**: Sabotage consists of the unlawful interception of communication traffic and invasion by a malicious third-party software, or as to interfere with the communication process. Disturbance of supply of products and services that failure due to shutdown may be caused by a variety of factors, including physical attacks, challenges to open CPS equipment, receiver types, and the whole electromagnetic infrastructure.

TABLE 4.2

Different Perspectives of Evaluating Cyber Security

Cyber security perspectives	Description
Centering information	Requires protecting the data flow during the storage phase, the transmission phase, and even the processing phase.
Oriented function	Requires integrating the cyber-physical components in the overall CPS.
Oriented threat	Impacts data confidentiality, integrity, availability, and accountability.

TABLE 4.3

Different Threats in CPSs

Wireless exploitation	It requires knowledge of the system's structure, and thus, exploiting its wireless capabilities to gain remote access or control over a system or possibly disrupt the system's operations. This causes collision and/or loss of control.
Jamming	In this case, attackers usually aim at changing the device's state.
Reconnaissance	This results in violating data confidentiality due to the limitation of traditional defenses.
Disclosure of information	Hackers can disclose any private/personal information through the interception of communication traffic using wireless hacking tools violating both privacy and confidentiality.
Unauthorized access	Attackers try to gain an unauthorized access through either a logical or physical network breach and to retrieve important data, leading to a privacy breach.
Interception	Hackers can intercept private conversations through the exploitation of already existing or new vulnerabilities leading to another type of privacy and confidentiality breach.
GPS exploitation	Hackers can track a device or even a car by exploiting (GPS) navigation systems, resulting in a location privacy violation.
Information gathering	Software manufacturers covertly gather files and audit logs stored on any given device in order to sell this huge amount of personal information for marketing and commercial purposes in an illegal manner.

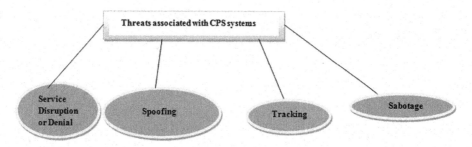

FIGURE 4.3
Threats associated with CPSs.

3. **Service disruption or denial**: Uninterrupted or error-free, that the store will attack the physical can affect the device to the break of work, the services, or to modify the configuration. These are serious consequences, especially in the case of medical applications.

4. **Tracking**: Due to the nature of the device is a physical person can receive in this unit, and/or even their age (Figure 4.3).

4.5.1.2 Physical Threats

Recently developed industries, the emergence of Innovative Metering Infrastructure (s), and the Country Network (Finance), as well as the storage yards of the monitoring and management, the dependability of the CPS, type of receiving antenna to the branch offices [13] all contribute to the widespread adoption of the CPS-type receiver. In fact, the natural hazards can be classified into the following three main factors:

4.5.1.2.1 Physical Damage

A variety of systems like the power grid, power station, base station are well-operated and well-protected, to implement a security mechanism that is based on access control, authorization, and authentication mechanisms, for example, the username and password, and maps, in the field of biometrics and video surveillance. Stations, which are the power lines, are the target of sabotage, attacks, and disruptions. The smart meter, and the delicate hot-sensors [11]. It's hard to get involved with the smart meter. Smart meters need to be constantly hacking the fullest possible information. Physical disruption of the peace and larceny on the side of the combatants, and not only to carry out. There are ways to lessen the blow of the danger; however, the impact of the risk can be reduced.

4.5.1.2.2 Loss

Aggressive, toxic threat in the event of a power substation failure is that there is a possibility of a complete shutdown of the large major cities, and within a few hours, the intelligent network is severely degraded [12].

4.5.1.2.3 Repair

The repair may be based on a self-healing process [14]. A self-healing process, which is based on the strength or sensation, interference with, or failure to isolate the issue and to alert the appropriate system to auto-configure the backup resources in a continuous supply of the required maintenance. In order to facilitate a speedy restoration, it is preferable that no time frame be specified. However, the critical components suffer from a lack of or a limited backup capability. Therefore, self-healing is able to react quickly to a serious injury.

4.5.2 CPS Vulnerabilities

In this section, we are going to present you the most important vulnerabilities of the CPS receiver, which may be the subject of the above-mentioned threats. Frankly speaking, it refers to the inability to fight against the enemies of the environment.

Analysis involves the identification and analysis of the weak points of the system that make the use of the CPS receiver. It also includes the identification of appropriate corrective and preventive measures to reduce, mitigate, or eliminate any potential vulnerabilities [15].

Table 4.4 gives a brief description of the vulnerability of the CPS receiver.

4.5.2.1 Reasons of Vulnerabilities

The vulnerabilities occur for many reasons:

Assumption and isolation: Probably also in the system, the challenge is to create a reliable and secure operating system, with a view for the implementation of any necessary service, provided that there is a system in isolation from the rest of the world.

Increasing connectivity: Increasing the connection, the receiver, as the CPS is now more connected to each other, and there is no guarantee that the software on these devices is a free vulnerabilities that could be exploited by a hacker. In most of the cases, the users and manufacturers address the CPS-type receiver connectivity, which is "install and forget" devices, which makes the potential security vulnerabilities in software accessible to the internet for many, many years to come. This is a type of risk which increases the risk that the system will have to be a victim to safety and security breaches.

Heterogeneity: Heterogeneity of the components of the third-party sites is cooked and presented in the CPS, the type of recipients for the creation of applications,

TABLE 4.4

Brief Description of CPS Vulnerabilities

Category	Description	Examples
Network vulnerabilities	Include weaknesses of the protective security measures, compromising open wired/wireless communication and connections.	Man-in-the-middle, eavesdropping, replay, sniffing, spoofing and communication-stack, back-doors, DoS/DDoS and packet manipulation attacks [16].
Platform vulnerabilities	Include hardware, software, configuration, and database vulnerabilities [17].	
Management vulnerabilities	Include lack of security guidelines, procedures, and policies.	

the CPS, the type of recipients. This diversity, in the end, it will increase the attack surface of the smart grid, and possibly increase vulnerability. Each of the heterogeneity of the components is subject to various security issues.

USB usage: Usage of USB is the main reason for the vulnerability of the CPS receiver. Anyone who knowingly distributes USB flash drives infected with malware that monitors a user's file system and network activity. Any gamer using a CPS system is vulnerable to the infection. Consequently, the thought, in and to the USB port. Then it was wired into a hub, from which it branched out to other devices for usage and duplication.

Bad practice: A bad experience in the first place because of the lack of encryption/ knowledge, which causes the execution of a code in an endless loop will be very easy to modify to an attack.

Spying: The receiver, which the CPS is under the attack of the spying/espionage, primarily, the use of spyware (malware) forms secretly enter and remain undetected for many years, and it is the most important task for the interception, theft, and a collection of proprietary/confidential information.

Homogeneity: For example, this is the kind of cyber physical systems, the suffering, alone, and at the same vulnerabilities, that is, when they are abused, it can affect all of the devices in its vicinity.

Suspicious employees: Suspicious-employed, they can be intentional or unintentional loss or damage to the receiver, which is a CPS unit, acts of sabotage, and to change the encoding of the language, or to provide remote access to hackers, to open a closed port, or attach an infected USB drive.

4.6 Generalized Attacker and Attack Models for Cyber Physical Systems

Many of the attack models are available for cyber physical systems (CPS receivers). An attack model is used to create a parameter-attack of the procedures and functions that are targeted to specific CPS receivers. A model which looks to be on the attack could capture both the physical and the loose, and it combines a number of existing attack models for a single structure.

Cyber physical systems (CPS for the recipient), the integration of computational algorithms and physical company attacker models are proposed for cyber physical systems (CPS). The attack models derived from the attacker model are used to generate parameterized attack procedures and functions that target a specific CPS. The attacker models can capture both physical and cyberattacks and unify a number of existing attack models into a common framework.

Cyber physical systems (CPS) are an integration of computational algorithms and physical components. CPS systems find their application in infrastructure like water treatment, smart grid, and transportation, and in smaller systems such as a pacemaker or an insulin pump. There is a possibility of cyberattack in such systems. It is important to design mechanisms for detecting and defending a CPS attack. Attacks may be cyber or physical. A "cyberattack" affects communication networks whereas a "physical attack" disturbs physical elements of a CPS such as motors, generators, pumps, etc.

TABLE 4.5

Various Types of Attack Models

Attack models	Attacks occurring
Network-centric models	Man-in-the-middle and denial of service attacks
Information security-centric models	Buffer overflow and code injection
Cryptography centric attacks	Plaintext, password cracking
CPS-centric attacks	Bias and surge attacks

There are different types of attacks that CPS systems have to defend against. Existing attack models can be used for defending against an attack or a new one can also be designed. Various of attack models are shown in Table 4.5.

The receiver system is the CPS that is used for infrastructure, water treatment, smart networks, and transportation, and in smaller systems, such as a pacemaker or an insulin pump. You will have the possibility of cyberattacks in such a system. It is essential to develop detection and protection mechanisms to prevent the attack on the receiver, as the CPS. The attacks could be a cyber or physical. In a "cyberattack," there is an impact on the network, and in this context, a "physical attack" as a violation of the physical components of a CPU, such as motors, generators, pumps, and so on.

To whom does the CPS-type technology need to provide protection, and against what kinds of attacks? Defending against an assault is possible with the modeling at your disposal, and you can even utilize these models to create new defenses (Table 4.5).

Various attack models are indicated in Table 4.5.

4.6.1 Limitations of Cyberattack Models

Cyberattack models suffer from the following limitations:

(a) Any non-compliance with the various types of attacks in a general-purpose system.

(b) No liability to the physical device.

These limitations can be overcome by careful design and method.

4.7 Security and Privacy in CPSs

We will discuss a wide range of applications in the industrial and personal CC of the CPSU and of the consequences of accident, protection, safety, security, and privacy. In a CPS-type receiver, the system shows many relationships, dependencies, and system interactions. The receiver is vulnerable to both digital and physical threats because of their dependency on the CPS. Additionally, the complexity of a physical system rises when its constituent parts are encouraged to communicate with one another via the use of a network.

The CPS-type of the receiver of the attack may be made to the cyber-physical-or domain-specific and domain, it is possible to be, the individual and the individual's protected, but it is only in one part of the system, where both of them at the same time, the domain (s) that is to be protected, is indeed to be protected.

These CPSs can have a variety of points of attack, security, and privacy, such as:

- The interface between the devices
- To your device
- The infrastructure that supports them
- The internet, as well as the
- Malicious users

4.7.1 Privacy in Cyber Physical Systems

Understanding the CPS's security requires an understanding of both the physical and cyber domains, and their interactions. The CPS's integrity appear to be more complex. When the findings are known, solid, receiver type, the CPS, and their interactions, then secrecy may be maintained.

In the internet age, all of the data collected is finally split up in a row. Already in the IoT the management of information is done by the computer manually. Now, medical devices, the state-of-the-art facilities, according to the drive, the cell phones, smart devices, connected cars, connected homes, and many other devices that are to be collected for an unknown amount and type of information to the users. Users are aware that these devices to interact over the internet. People fail to realize is that their devices are connected to the internet, and they do not fully understand the implications of being safe and secure. This can result in the leak of information that could be gathered, or stolen, without the knowledge of the affected user.

4.7.2 Security in Cyber Physical Systems

This example is imported in this section to say that the safety of the complexity of a system is to make the transition from cyber or physical to cyber-physical. We'll talk about the CPSU central committee's physical setup and its people, as well as the committee's overall contemporary design in Figure 4.4.

4.8 Conclusion

In this chapter, we have defined the security and integrity, the application of the classical definition, both in the digital and physical worlds, and the development in the area of the CPSs. We have illustrated the various security and privacy issues between the infrastructure and the individual CPSs and shown how the communication between the systems is affected, that is to say, the need for security and integrity of the network infrastructure. A number of examples showing what happens when a part of the security and the integrity is neglected have been presented, and we have also discussed the methodology in order to ensure that the system has been designed and has been used in a wrong way. Finally, we discuss the current challenges faced by consumers and the industry, as everywhere, the spread of the CPSs is on its way to becoming the new norm.

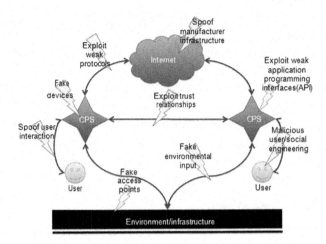

FIGURE 4.4
Various attack points in CPSs.

That is, as the number of devices that are connected to the computer, such as the CPSs, is continuing to grow at an exponential rate, safety, security, and privacy needs to be taken seriously. Attention should be given to both and the physical dimensions of the CPSS, as well as their interaction. Also, there is a serious concern, and the risk of substance abuse can be decreased by the existence of these devices is on the formulation of the project, or the general arrangement is adopted, the costs and benefits should be considered. The central committee of the CPSU, is it forever, and their presence is going to have a greater impact on the daily lives of billions of people around the world. So, the security and the privacy of individuals is incalculably important in the development, promotion, management, operation, and replacement of the CPSs.

References

1. Lee, J., Bagheri, B., & Kao, H. A. (2015b). A Cyber-Physical Systems Architecture for Industry 4.0-Based Manufacturing Systems. *Manufacturing Letters*, 3, 18–23. https://doi.org/10.1016/j.mfglet.2014.12.001
2. Lu, Y. (2017b). Industry 4.0: A Survey on Technologies, Applications and Open Research Issues. *Journal of Industrial Information Integration*, 6, 1–10. https://doi.org/10.1016/j.jii.2017.04.005
3. Lee, J., Lapira, E., Yang, S., & Kao, A. (2013b). Predictive Manufacturing System - Trends of Next-Generation Production Systems. *IFAC Proceedings Volumes*, 46(7), 150–156. https://doi.org/10.3182/20130522-3-br-4036.00107
4. Heeg, R. (2014). The Future is Wide Open: Stefan Lindegaard on Open Innovation. *Research World*, 46(46), 57–60. https://doi.org/10.1002/rwm3.20105
5. Lee, E. K., Viswanathan, H., & Pompili, D. (2015). Distributed Data-Centric Adaptive Sampling for Cyber-Physical Systems. *ACM Transactions on Autonomous and Adaptive Systems*, 9(4), 1–27. https://doi.org/10.1145/2644820
6. Kang, S., Chun, I., Park, J., & Kim, W. (2012b). Model-Based Autonomic Computing Framework for Cyber-Physical Systems. *IEMEK Journal of Embedded Systems and Applications*, 7(5), 267–275. https://doi.org/10.14372/iemek.2012.7.5.267

7. Rad, C. R., Hancu, O., Takacs, I. A., & Olteanu, G. (2015b). Smart Monitoring of Potato Crop: A Cyber-Physical System Architecture Model in the Field of Precision Agriculture. *Agriculture and Agricultural Science Procedia*, 6, 73–79. https://doi.org/10.1016/j.aaspro.2015.08.041

8. Zahid, F., Tanveer, A., Kuo, M. M. Y., & Sinha, R. (2021). A Systematic Mapping of Semi- Formal and Formal Methods in Requirements Engineering of Industrial Cyber-Physical Systems. *Journal of Intelligent Manufacturing*. Published. https://doi.org/10.1007/s10845-021-01753-8

9. Chaganti, R., Gupta, D., & Vemprala, N. (2021). Intelligent Network Layer for Cyber-Physical Systems Security. *International Journal of Smart Security Technologies*, 8(2), 42–58. https://doi.org/10.4018/ijsst.2021070103

10. Yaacoub, J. P. A., Noura, M., Noura, H. N., Salman, O., Yaacoub, E., Couturier, R., & Chehab, A. (2020b). Securing Internet of Medical Things Systems: Limitations, Issues and Recommendations. *Future Generation Computer Systems*, 105, 581–606. https://doi.org/10.1016/j.future.2019.12.028

11. Mahmoud, M. S., Hamdan, M. M., & Baroudi, U. A. (2020). Secure Control of Cyber Physical Systems Subject to Stochastic Distributed DoS and Deception Attacks. *International Journal of Systems Science*, 51(9), 1653–1668. https://doi.org/10.1080/00207721.2020.1772402

12. Xu, Y., & Singh, C. (2014). Power System Reliability Impact of Energy Storage Integration with Intelligent Operation Strategy. *IEEE Transactions on Smart Grid*, 5(2), 1129–1137. https://doi.org/10.1109/tsg.2013.2278482

13. Maknouninejad, A., & Qu, Z. (2014). Realizing Unified Microgrid Voltage Profile and Loss Minimization: A Cooperative Distributed Optimization and Control Approach. *IEEE Transactions on Smart Grid*, 5(4), 1621–1630. https://doi.org/10.1109/tsg.2014.2308541

14. Chopade, P., & Bikdash, M. (2016b). New Centrality Measures for Assessing Smart Grid Vulnerabilities and Predicting Brownouts and Blackouts. *International Journal of Critical Infrastructure Protection*, 12, 29–45. https://doi.org/10.1016/j.ijcip.2015.12.001

15. Amin, S., Litrico, X., Sastry, S., & Bayen, A. M. (2013b). Cyber Security of Water SCADA Systems— Part I: Analysis and Experimentation of Stealthy Deception Attacks. *IEEE Transactions on Control Systems Technology*, 21(5), 1963–1970. https://doi.org/10.1109/tcst.2012.2211873

16. Venkatraman, S., & Venkatraman, R. (2019). Big Data Security Challenges and Strategies. *AIMS Mathematics*, 4(3), 860–879. https://doi.org/10.3934/math.2019.3.860

17. Adepu, S., & Mathur, A. (2021c). Distributed Attack Detection in a Water Treatment Plant: Method and Case Study. *IEEE Transactions on Dependable and Secure Computing*, 18(1), 86–99. https://doi.org/10.1109/tdsc.2018.2875008

18. Palleti, V. R., Adepu, S., Mishra, V. K., & Mathur, A. (2021). Cascading Effects of Cyber-Attacks on Interconnected Critical Infrastructure. *Cybersecurity*, 4(1). https://doi.org/10.1186/s42400-021-00071-z

19. Adepu, S., Shrivastava, S., & Mathur, A. (2016). Argus: An Orthogonal Defense Framework to Protect Public Infrastructure against Cyber-Physical Attacks. *IEEE Internet Computing*, 20(5), 38–45. https://doi.org/10.1109/mic.2016.104

20. Hahn, A., Ashok, A., Sridhar, S., & Govindarasu, M. (2013). Cyber-Physical Security Testbeds: Architecture, Application, and Evaluation for Smart Grid. *IEEE Transactions on Smart Grid*, 4(2), 847–855. https://doi.org/10.1109/tsg.2012.2226919

21. Franze, G., Tedesco, F., & Lucia, W. (2019b). Resilient Control for Cyber-Physical Systems Subject to Replay Attacks. *IEEE Control Systems Letters*, 3(4), 984–989. https://doi.org/10.1109/lcsys.2019.2920507

22. Liu, Y., Ning, P., & Reiter, M. K. (2011). False Data Injection Attacks against State Estimation in Electric Power Grids. *ACM Transactions on Information and System Security*, 14(1), 1–33. https://doi.org/10.1145/1952982.1952995

23. Lee, J., Bagheri, B., & Kao, H. A. (2015c). A Cyber-Physical Systems Architecture for Industry 4.0-Based Manufacturing Systems. *Manufacturing Letters*, 3, 18–23. https://doi.org/10.1016/j.mfglet.2014.12.001

24. Lu, Y. (2017b). Industry 4.0: A Survey on Technologies, Applications and Open Research Issues. *Journal of Industrial Information Integration*, 6, 1–10. https://doi.org/10.1016/j.jii.2017.04.005

25. Lee, J., Lapira, E., Yang, S., & Kao, A. (2013c). Predictive Manufacturing System - Trends of Next-Generation Production Systems. *IFAC Proceedings Volumes*, 46(7), 150–156. https://doi.org /10.3182/20130522-3-br-4036.00107

26. Gao, L. S., & Iyer, B. (2009). Value Creation Using Alliances within the Software Industry. *Electronic Commerce Research and Applications*, 8(6), 280–290. https://doi.org/10.1016/j.elerap .2009.04.009

27. Wang, P., Ma, M., & Chu, C. H. (2018). Long-Term Event Processing over Data Streams in Cyber-Physical Systems. *ACM Transactions on Cyber-Physical Systems*, 2(2), 1–23. https://doi.org/10.1145 /3204412

28. Kang, S., Chun, I., Park, J., & Kim, W. (2012c). Model-Based Autonomic Computing Framework for Cyber-Physical Systems. *IEMEK Journal of Embedded Systems and Applications*, 7(5), 267–275. https://doi.org/10.14372/iemek.2012.7.5.267

29. Rad, C. R., Hancu, O., Takacs, I. A., & Olteanu, G. (2015c). Smart Monitoring of Potato Crop: A Cyber-Physical System Architecture Model in the Field of Precision Agriculture. *Agriculture and Agricultural Science Procedia*, 6, 73–79. https://doi.org/10.1016/j.aaspro.2015.08.041

30. Venmani, D., Lagadec, Y., Lemoult, O., & Deletre, F. (2018). Phase and Time Synchronization for 5G C-RAN: Requirements, Design Challenges and Recent Advances in Standardization. *EAI Endorsed Transactions on Industrial Networks and Intelligent Systems*, 5(15), 155238. https://doi.org /10.4108/eai.27-6-2018.155238

31. Gunduz, M. Z., & Das, R. (2020b). Cyber-Security on Smart Grid: Threats and Potential Solutions. *Computer Networks*, 169, 107094. https://doi.org/10.1016/j.comnet.2019.107094

32. Yaacoub, J. P. A., Noura, M., Noura, H. N., Salman, O., Yaacoub, E., Couturier, R., & Chehab, A. (2020c). Securing Internet of Medical Things Systems: Limitations, Issues and Recommendations. *Future Generation Computer Systems*, 105, 581–606. https://doi.org/10.1016/j .future.2019.12.028

5

Challenges in the Taxonomy of Cyber-Physical Security and Trust

Piyush Parkash, Nitin, and Shalini Singh

CONTENTS

5.1 What Is Security and Trust?

- Security violations are reported almost every day. It is enticing to use its supposed identity to further the argument for security and reliability. Over the past couple of years, we have seen Facebook's role in huge data collection [1], personal data breaches from Equifax [2], and personal voice-activated companions [3] being

DOI: 10.1201/9781003220664-5

deported. Although it is true today that every system designer would do well to examine the safety ramifications of their work, we also emphasize additional important reasons for considering cyber-physical microfluidic biochip (CPMB) security:Untrusted supply networks: Transnational distribution channels in the contemporary production of semiconductor products result in a complex web of players with diverse capacity and motive for harmful actions such as hardware insertion from Trojan [4] and piracy for property rights (IP) [5]. Commercial microfluidic devices of the next generations should adapt to the sophistication of manufacturing processes and will therefore be subject to the same assaults [6].

- In addition to being a boon to developing secure and reliable systems, cyber-physical integration also poses a severe safety problem for a number of reasons. Firstly, it involves the presence of a digital control unit. This controller usually is built with a single circuit board or an off-shelf microcontroller and can add several new potential vulnerabilities such as a network interface. Second, an intruder may exploit the physical features of the system to tremendous advantage; the data is not just the only thing in danger, but physical properties can also now be altered and destroyed. By exploiting the app's physical weaknesses, scientists may learn more about the application's inner workings and exposing it to possible threats.

- Increased adoption rate: Microfluidics literature has repeatedly shown that this technique still has to be marketed on a wide scale [7]. Security and confidence are elements that may be developed and automatically converted into CPMBs and which provide a high degree of protection. In contrast to macrolabs, training conformity with regulations and audits is the norm under human supervision. The security-centric design might eventually lead to viable solutions for general implementation using microfluidics from mostly academic studies.

- Prevention: After the fact, the computing sector is seeing the adoption of safety measures. A response-oriented strategy guarantees that genuine issues prompt solutions but at a significant expense for the people involved. Microfluidics is still a new technology that offers few motivating examples. But this is essentially why microfluidics provides an opportunity since the architecture of safety and confidence can first and foremost prevent such events.

- Amazing: Microfluidic technology can be pretty convenient for safety-centered design. The designers can construct security networks naturally using unique microfluidic characteristics. For example, in general, the complexities of technologies such as these are smaller than that in VLSI [8], and time scales are frequently less than those in the semiconductors in microfluidic systems [9]. In addition, standards have to be released as scientists continue to investigate new notions of design and study fundamental physics and materials [7]. Thus, there is a chance to include safety in standards development.

5.2 Classification of Attacks

Here, we present an integral classification of potential vulnerabilities, threat models, and motives that may affect CPMBs. We notice that while hypothetical, this categorization is based entirely on ideas and instances from other fields. The categorization emphasizes elements that are unique to microfluidics as much as practicable.

5.2.1 Attacking Area

- A surface for an assault is a possible entry point for an offensive attack. According to the following taxonomies, the attacking regions can have:
- Indirect physical access: unintentional usage of IP address and port number to perform an attack. This would include uncovered USB and Internet jackets initially used to maintain or transfer information and the unloading of cartridge trays for biochips. In addition, some microfluidic systems are built on ordinary, in-shelf processors and accessible to a variety of underutilized physical ports.
- Direct physics access: Many developers aim to develop a fully portable, autonomous lab-on-one chip. This offers numerous practical advantages but physically renders the platform insecure. For example, a gadget for pollution management installed in remote areas would readily be manipulated. Fault injecting through power or clock smoking and exploiting controllers, samples, or chemicals may be feasible in intrusive, direct physical attacks, in particular, in susceptible environments. This also implies the privacy of wearable sensors, as signal tracks may be picked off.
- Network access: Network-enabled cyber-physical connectivity offers a robust assassination surface for malevolent opponents seeking distant attacks. Zero-day feats, intelligence agency hacking tools, uncached systems, and social technology contribute to networked computer vulnerability. It complicates the reality that CPMBs are typically implemented in environments that cannot be maintained regularly either because of costs or inaccessibility. The only method to avoid network interfaces being misused is to eliminate them, but the utility of the CPMB would be significantly limited.
- Wireless access: The convenient way to transfer data, especially to smartphone applications integrated with mobile terminals, such as wirelessly and using Zigbee. This offers the possibility for a microfluidic platform for attackers nearby but not necessarily in possession.
- Documentation of the configuration: The knowledge needed to make a foundation for the microfluid might consist of circuitry schemes, procedures for biochemical testing, or biochip layouts. System specification may be easily exploited for adding capacity attacks in their raw form.
- Default interfaces: This refers to user experience, such as a touchpad or a chassis cut-out, which is often required to observe the advancement of the test. If not correctly constructed, such connections may disclose protected information or allow permissions to escalate. For example, damages may ensue if operational parameters that the user may select are not safe from unauthorized or hazardous settings. Or, by carefully constructing a collection of liquids that can signal the blending sequence, an attacker may try to revise a protocol.
- Side-channel attacks utilize physical phenomena that were initially not considered in system design. Insecurity literature, microwave radiation, and power side channels were primarily investigated and proven to be effective in disrupting unprotected cryptography algorithm hardware implementations [10]. CPMBs

have different physical properties, such as temperatures, luminescence, and metabolites, that can be used for this purpose.

5.2.2 IP Attack Modeling

- Based on the criteria provided forth in, we classify an enemy country tactical and operational capability [11]. Technical skills represent an attacker's understanding of how the microfluidic platform operates and its ability to obtain this experimentally based intelligence. Operations skills define the way an opponent may use or inject the type of malware, for example, I/O ports on a microflow platform, while an IP assault implies the assailant has access to an installation. Note that because we include IP-based attacks, our concepts of security and reliability are more comprehensive than those mentioned in [11]. We offer possible threat models categorized by attacker sites for researchers to determine.

- The untrusted supply chain is responsible for generating threat models. Biochip platform developers should deal with components from different suppliers and combine them. These suppliers might be situated abroad, and many suppliers can be employed simultaneously. These parties would probably wish to carry out IP-related assaults and have considerable technical and operational skills since vital design information is supplied.

- Once the microfluidic technology is implemented and functioning, field risks occur. Opponents might include malevolent end users wishing to change the operation of a technology and distant adversaries involved in data or natural facilities being compromised. Those opponents can have excellent technological skills, especially remotely parties, which can be found globally and supported by national governments. They have limited operating skills, as their ability to attack is constrained by the accessible equipment and software attack vectors.

5.2.3 Difficult Prediction

There is a wide range and challenging prediction of human incentives to compromise cyber physical systems; however, based on various relevant disciplines, we may list several typical motives that are likely to drive the assault.

- Financial gain: All IP assaults are motivated. There is plenty of evidence that IC products are counterfeit, over the manufacture, and reverse engineering. Other assaults may be driven by malware, ID theft, and sales of user data. Intriguing and recent phenomena are unethical marketing attacks, which can reveal a security weakness and lead to significant changes in the market value of a firm. This has occurred recently with the surprising revelation of numerous vulnerabilities in AMD processors [12].

- Revenge: Disgruntled workers have reported some of the most prominent safety violations in recent years [13]. These assailants are incorporated in an organization, and after they have been incorrect, they use the destructive effect from their accessibility and expertise.

- Policy: The Stuxnet worm's emergence in 2010 reassessed the actual scale of government-sponsored, politically driven cyber threats [14]. Engineers working on cyber-physical systems should consider formidable adversaries who could steal trade secrets or conduct other forms of damage to the system for personal gain.

- Personal gain: Academics may be motivated to manufacture data under pressure to publish. Dubious scholars may be encouraged to look for advanced fraud tactics due to the rising efficiency of the procedures used to identify falsified data.

5.3 Intrusion Outcomes

- Due to this flow of architecture and the typical biochip building, many kinds of assaults are conceivable. Depending on the success, we may generally rate attacks as follows.

5.3.1 Modifying Sensors

- Reading falsification refers to the modification or manufacturing sensor readings to deceive the end user. After digesting and sensing the microfluidic platform, the computer that eventually stores, analyzes, and sends sensor information can be interfered with. The use of message security mechanisms [15] can handle computer-based assaults, but cyber-physical interconnectivity leaves the potential of a physical assault.

- As microfluidics finally moves toward wearable electronics that may be quickly implemented in the field, the physical aspect becomes more and more a worry. A hostile end user can modify the sensor equipment by disrupting the sensor/controller link. A fake picture can be presented before the camera system in the optical detecting systems to delude the controllers. This might lead to a medical diagnosis application's harmful or disastrous patient results. Alternatively, an adversary might change sensor values for errors detection equipment used on the biochip.

5.3.2 Attack on Services

- Denial-of-service (DoS) is a type of attack that breaches the system's unavailability and makes it useless. DoS is experienced more often as a nuisance on the Internet every day. There is a possibility for significantly more damage in a microfluidic biochip. It might be costly to collect and refill specimens that are suitable for use with microfluidic devices. Several hardware defects such as external contamination make dependable functioning difficult [16]. DoS is hard to protect against since an intruder must only exploit a single vulnerability, and an operation can be carried out at any stage of the design flux. The dynamics that make DoS prevalent nowadays for computer security, such as computer platform homogenization and DoS-as-a-service, will likewise play an unwanted role in microfluidic biochip deployments if no attention is paid to this design stage.

5.3.3 Altering Function

- Functionality changes are a danger to the undesired performance of a CPMB. The line between denial of communication and functioning is complex. Aside from the fact that the device is now inoperable, it also continues to perform in an unanticipated or degraded manner. The device is now functioning. A product that fails to comply with specifications may cause frustrations for the end user and may cause the producer to lose customers. We see abuse from the end user as a viable place to target functional modifications; to attain an intended result, if an end user wants to bend the microfluidic platform to him, it is a breach of the authenticity of the assay. It violates the confidence in the applications in which the platforms are implemented on a broader scale. A hostile end user poses a severe concern because it is challenging to construct manipulative devices caused by physically intrusive attacks [17].

5.3.4 Designing Robbery

- Design robbery is a wide range of copyrighted material and cyber-physical device-manufacturing risks and knowledge. The intellectual capital in the microfluidic realm exists as masks, substances, and procedures that form a substratum for manipulating fluids and protocols. These are thus instructions for the performance of the biochemical analysis. For the most part, sophisticated technology is required to develop these protocols, and businesses want to safeguard their funds by preventing unauthorized use. In general, in 2014, the market of LOC technologies reached $3.9 trillion and is predicted to rise by $18.4 trillion by 2020 [18]. Design robbery might increase the sector's significant growth in future years. Arguments over property rights infrequently happen in microfluidics [19] and waste the financial capabilities that might be better used somewhere. In addition, when the offenders of invention theft are situated overseas, there may be the minimum legal consequence. Halbleiter suffers from production constraints such as over construction, IP theft, and counterfeiting [20]. It is fair to expect the microfluid sector to face difficulties because many manufacturing techniques are developed from semicontroller procedures. With the increasing complexities of microfluidics, more vertical integration of the supply chain is expected. The reliability of the design flow [6, 21] thus leaves unanswered concerns.
- IP theft can even happen at the end of end-user fluid operation and is commonly seen. A standard microfluidic device is produced using a transparent plate, as an image sensor is required to detect the end output. The motion of fluids on the device is thus easy, and the procedure based on these measurements is reverse engineered. Low complexity and low operating characteristics aggravate the issue, and thousands of hertz are typically operated by digital microfluidic biochips (DMFBs) [22].

5.3.5 Leakage of Sensitive Data

The unlawful disclosure of personal or confidential data is a security and privacy issue. Untrustworthy supply chains provide numerous avenues of attack for knowledge leaking. Patient data, unique biochemical procedures, and private keys are examples of personal knowledge. Many microfluidic biochips are used in medical settings for diagnosis. Therefore, critical patient information needs to be managed appropriately. Due to the nature of the hardware industry's supply chain, horizontal research has been given a high priority. Trojans were shown to be compact despite divulging sensitive data, including cryptographic private keys. Since most of the ingredients and procedures used for manufacturing microfluidic devices immediately borrowed from the IC sector are also relevant, microfluidics' peculiar physics can lead to susceptible communication systems.

5.4 Malicious Modification in Signals

A fraudulent change to impulses used to operate a biochip is an act of manipulation attack initially described as part of DMFBs with changed actuation sequences [23]. The assault outlines how a hostile opponent might accomplish results such as denial-of-service or statistic manipulation. Various attack vectors may perform power transfer manipulation, such as data manipulation in memory locations, the programming alteration used to produce actuating sequencing or physical equipment insertion. Flow-based biochips can potentially be tampered with by their pressure control valve to actuate [24]. The overall efficacy and adaptability of manipulation assaults make it appealing to malevolent opponents. The book's hardware security approaches include counteracting assaults through actuators manipulation.

We are examining ways to mitigate digital polymerase chain reaction (dPCR), an important application field for microfluidics, to emphasize the risk presented by this assault in a CPMB [25].

5.4.1 Quantifying and Amplifying Nucleic Acids in a DNA Sample

A comparatively new approach for measuring and multiplying nuclear acids in template DNA is the digital polymerase chain reaction. The input must be divided into numerous tiny reaction chambers by volumes and is different from standard PCR methods. dPCR has several strengths, such as inhibitor tolerance, absence of calibration curves, and the opportunity to give quantitative actual, not comparative [26]. We discuss the ideas underlying dPCR in this section and how a commercial microfluidic device developed for dPCR may be hacked so that the dissemination of target DNA among some of the compartments is partial. Since a benchtop device contains the dPCR processes and physical valve processes, the end user would be unable to attack if the biochip was checked for statistical abnormalities and evaluated, thereby completely deregulating the goal of benchtop automating. We will then address the consequences for variance in copy numbers and the overall integrity of research.

5.4.2 Attacks on Multiple Small Reaction Chambers

The standard implementation of the microfluidic dPCR chip-based catalyst is used for the development. The reaction cell, including inlets for materials and reagents, is a detachable chip. The chip is installed into an embedded controller, which incorporates a variety of sensors. To produce numerous tiny reaction chambers, the microcontroller analyzes the on-chip actuators and cycles the temperatures to conduct PCR responses. Embedded fluorescent detectors communicate the test results to the built-in computer that displays or stores the data on an incorporated monitor. The operation of the microfluidic platform streamlines several previously performed manual procedures. Therefore, user mistake falls quickly. In this operations protocol, however, it is implied that the devices execute the test with integrity, as the end user does not take any step from the sample preparation to the final reading.

Suppose an attacker can disturb the activation of the microwave biochip. In that case, it may be possible to skew the dispersion of DNA samples or stop the PCR process, resulting in inaccurate estimations of the actual target DNA concentration. We have investigated a marketable microfluidic platform that has been determined to reflect its architecture, as illustrated accurately by an exposed attack surface on the USB, serial, and ethernet interfaces. An attacker can load computer viruses if these channels are unprotected. We found out that a single board computer with bespoke software loaded into a CompactFlash (CF) card is utilized on this specific platform off-the-shelf. The replacement of a CF card by a malignant one might jeopardize a technically susceptible machine. An external entity might use the connection to the internet to take over the device. An assailant might partially cause the pneumatic actuators to malfunction when the microcontroller is affected. The commands may be adjusted to shorten the start time or provide a transitional control signal. The opening of elastomer valves has a linear response to pressure changes over most of their operation [27]. Consequently, the sampling rate would be interrupted in all the reaction chambers, and a Gaussian system with constant parameters would be breached.

5.4.3 Operating Principles of dPCR

dPCR's working concept is to divide the DNA sample arbitrarily into many tiny chambers. These reaction chambers may be carried out in a chip-based array of dPCRs or created by encapsulation of materials in droplets produced by oil emulsion (droplet dPCR). The PCR amplification is done on all partitions to enhance the specific DNA sequence. The fluorescence monitor then reads the reaction. The percentage of affirmative to negative responses can be utilized in calculating the number of nucleic acid sequences in the sample. The aim is to have a Poisson distributive distribution following the randomized data division and determine the expected number of target nucleotide sequences (M').

$$\hat{M} = -\ln{(1 - H/C)}^{\hat{}} \qquad\qquad 5.1$$

Where H Ç is the affirmative chamber seen, and C is the total compartments [28]. Inconsistent receptacles and a non-random arrangement of molecules have, yet, restricted the adoption of dPCR for clinical diagnostics [29] in their assessment of positive chambers.

dPCR methods have been loaned to microfluidic technology, which provides applications like multiple genetic variability investigations and medicine processing. These devices are presently being offered only for research, and diagnostic applications are likely to occur only once the technology matures [26, 29].

5.4.4 Simulation Process in dPCR

A large-scale mathematical model demonstrates the effects of an assault. The random assignment of M molecules into C tetra reaction chambers dispersed through K of a panel can be mimicked by a dPCR experiment. In other words, we choose one of the reaction compartments and assign the component to each component with a probability proportional. We next evaluate the genuine concentrations of molecules based on observation of the variety of positive reaction chambers H and assume that the identification works correctly in these simulations.

5.4.5 Detection Capabilities

Differences in architectural repetition numbers (CNVs) in genomic segments are differences [30–32]. CNVs are different. CNVs have been researched by the sensitive and precise identification of techniques like dPCR, which have provided significant insight into how minor variants play a role in genetic disorders such as autism and Crohn's condition. Multiple genes ratios established during illness investigations may be affected by the quantity of positive responses an attacker generates while undermining microfluid frameworks for the performance of dPCR. Without the correct copy numbers, positive connections between these genomic changes and illnesses are not possible. Worse, it may be possible to create faulty connections. Spurious relationships will prevent therapies for improvement.

5.4.6 Fabricating Data

Instead of entirely falsifying data to make more compelling proof that the experiments have been performed, the attacker might be driven to interference with materials and equipment. Attacks on parts required threaten to overturn recent orders to improve research performance and repeatability. In particular, within the dPCR, many investigators were unaware of the technical, fundamental methods and pitfalls—starting from an initial was result published in 2013 expressly to address those questions for posting Mathematical Digital PCR Experimentation (Digital MIQE) guidelines [33]. However, an investigator might cooperate fully with the norm and still produce unreproducible data. Therefore, time and resources need to be squandered to discover these false outcomes.

Scientists or technicians in the laboratory are sometimes driven to create or strengthen financial data. These breaches of research excellence can significantly affect everyday individuals, apart from defecting scientific publications. Currently, dPCR is utilized in research settings because of the cost of the technology involved and resource needs. dPCR microfluidics technologies are predicted to mature to become an appealing diagnostic platform. Hence, a PCR platform assault's possible safety consequences may jeopardize patients' well-being. An attacker might affect a medical provider's decision-making by distorting the definitive diagnosis to the extent that they are within reach.

5.5 Challenges of Microfluidic Security

We now give thorough analyses on applications that present a substantial obstacle or a possibility for creative development to further stimulate work in microfluidic safety.

5.5.1 Privacy in Medical Diagnostics

Medical diagnosis constitutes an essential field of applications for research into microfluids. It will be advantageous for patient participation and better treatment quality to swiftly identify illnesses and multiplex tests in distant areas outside the medical centers. However, if precautionary steps are not implemented, the multiplication of untrusted inexpensive gadgets will result in many invasions of privacy. Data gathered after the processing of fluids and personal metadata might be included in confidential documents on a microfluidic device. Transparency of patient information has become a concern to medical equipment manufacturers [34]. Many of the same problems and risks will apply to microfluidic cyber physical systems.

The biological laboratory analytical chips would theoretically undergo approved disposal processes; however, regulations guaranteeing safe disposal will be divided between countries and the Environmental Protection Agency [35]. In addition, safety is not part of a routine. In health care facilities, in physically insecure plastic containers, the biohazard collecting containers are situated. Sterilization procedures, such as burning, autoclaving, and microwaving, are quite damaging and likely remove confidential material. However, these technologies have negative limitations such as energy, area, and expense for expensive equipment and pollutants generation. There are now no disposal restrictions for biochips, and there will probably be no time for any legislative changes. And although standards are publicized and mandated, diagnostic equipment that employs remote controls cannot rely on end users to follow proper disposal procedures. Designers must guarantee that critical information may be safeguarded even if no architectural safety measures are introduced when the device enters the appropriate disposal.

5.5.2 Capability of Public Safety

Microfluidics offers an excellent framework for public protection and defense applications. The enhanced water and wastewater treatment capacity and the benefit of a larger specific surface area ratio allow the analysis of chemicals to carry out an abundance of unusual reactions in the chip. Examples include environmental pollutants monitoring [36, 37], pathogen identification, food security protection medications [38], and chemical or biological weapons [39], for instance, microfluidic devices intended for public collision avoidance. Early identification of toxins and infections is one of the essential elements in successful protection against biological weapons. Early diagnosis allows therapy before symptoms occur entirely while also safeguarding against people who have affected immune function like the elderly. Detection techniques must thus operate reliably and efficiently at all times.

It is hard to estimate the actual threat of biological warfare as 170 nations have accessed and accepted the Convention on Chemical Weapons 1972 [40]. Whether these governments are now actively producing or hosting bioweapons is entirely speculative. But events in Damascus have demonstrated the development and deployment of biological weapons: the sarin nerve toxin was discharged in armed form. It was responsible for the slaughter and injury of people. The Assad regime refused to engage and called the tale manufacturing. Sustainable environmental surveillance devices allowing microfluidics might offer an early warning and an assault at a time when the fundamental facts are being discussed about actual occurrences.

The disability of detection techniques will also be necessary for organizations concerned about guaranteeing optimum performance and believable negligence of their biological warfare. Countries are one of the few entities that are able and adequately determined to create huge stocks of biological weaponry. Therefore, security designers have to presume that attackers are technically advanced.

Immunoassays and nucleic acid screening are two approaches used to identify biological and chemical weapons on the chip [39]. Immunoassays detect how much analytic a solution has by attaching an antibody to a particular macromolecule. At the same time, PCR is used to enhance and recognize pathogens of DNA in nucleic acid tests. For all types of detection techniques, several DMFB-based devices were mentioned in the documentation. The manipulation might be done using software or by loading Trojan hardware. In biochips with continuous flow, gates are computerized and may thus be tampered with the same action. Then fluids can mix or produce early insufficiencies accidentally.

5.5.3 Basic Research Integrity Tool

In the chemical and biological disciplines, microfluidics as a fundamental research tool helps to increase performance and automate complicated processes that enhance reproductivity and dependability while decreasing operating mistakes. Indeed, microfluidic methods were predicted to replace the bottle. Despite these upsides, microfluid studies in engineering vs. biology and healthcare have a primary connection. The more productive engineering microfluidics publications are devoted to the proof-of-conception of new technologies and techniques. Biological and medical researchers adopt these technologies slowly and widely, while microfluidics remains limited. Instead of just reproducing current laboratory procedures, research has shown that microfluidics researchers might profit from inventing new methods and applications. One such use might be confidentiality and trustworthiness.

As demonstrated in the case study of dPCR, microfluidic systems might give an advanced path for false investigators to commit investigational fraud credibly. Although peer review and findings can automatically screen for fraud, the procedures are reactive. This leads to situations like the disgraced scientist Haruko Obokata whose mentor died suddenly after retracting her Nature publications when scientific fraud results in a human loss.

The second fraud incident occurred in 2010 when a third-party research facility entrusted with FDA medication assessment mainly produced data over the years. This information was utilized for medication approval, with about 100 allowed on the marketplace. And despite this disclosure, many of these medicines remain on the market for pharma manufacturers. The problem might potentially have been averted if the instruments utilized for these experiments had provided safe and trustworthy microfluidic technology.

Can any sort of integrated error checking used in mass spectrometers prevent such high levels of fraud from occurring? Another strong justification for preventive actions was made in the FASEB publication: "The longer the scientific community responds to events that reduce confidence, the more it leaves the public luck to fund its work." Ambitious aims, such as maintaining the integrity of research and certifying sensor readings, are functions that conventional technical laboratories cannot achieve; the advances in safety and the confirmation of confidence clearly illustrate the superiority of laboratories. The equipment and machinery used to carry out the research may be controlled using cyber-physical microfluidics to ensure responsibility. To safeguard researchers' privacy, it would only be possible for service providers to examine the information for this study at issue. Microfluidics is appealing in this respect as a framework for doing research—it allows accurate fluid handling while incorporating sensors to track experiential stages in real time. Developments in condensed sensing might also help effectively capture the result of an investigation.

5.5.4 Blood Testing Platform

Monsanto has developed Edison, a technology for microfluidic blood tests, in the mid-2010s. Only a few droplets of blood were needed while the prices were reduced to really cheap levels and was seen as innovative. However, the firm received criticism and unfavorable media over their cloud computing scientific basis. In 2016, when Monsanto had to record blood test data for several years, the criticism appeared well-founded. Consumers were confronted that erroneous laboratory tests determined their medical attention. The technical specifics of the case of Monsanto are not known, but the compromising of a well-used blood diagnosis test may be investigated: in vitro glucose measurement. Regular diabetes testing is required for appropriate monitoring of diabetic individuals. The amount of insulin to be administered to the patient is decided dependent on the extent of glucose levels. The smoothly conducted nightstand glucose tests with automatic microfluidic biochips provide fast, low-cost, and effective measurement. A diagram represents the glucose-measuring test, known as a sequenced network. This test usually evaluates the amount of glucose in a blood sample by building a known concentration solution's inoculating loop of the glucose calibration curve. The X-axis shows the various concentrations of dilution (mg/dL), and the Y-axis is the response rate measured by the change in the absorbance grade expressed as AU/s (absorbance unit per second). This curve helps to interpolate the concentrations of the glucose sample being tested, which is why it is necessary to achieve maximum accuracy.

Two realistic result-manipulation assaults against glucose-based microfluidics have been disclosed in Ali et al. The first attack (Attack 1) wants to mess with the test results by altering the diabetic content in the sample. The heavy dotted lines represent sequence graph modifications relative to the brilliant sequence graphs. The S3 buffer droplet W1 is combined with the I6 glucose outlet and diluted Dl10. Since hypoglycemia is halved in intensity, the test result is incorrect. The user cannot understand the absorbance values of the waste buffer droplets. A malformed calibration curve initiates the second attack (Attack 2) by manipulating the sequence graphs on the reaction chains 2 and 4. For this reason, the two D1 and S3 waste buffer droplets are utilized.

Intentionally destructive attack network was created by malicious attackers. The thick lines represent the changes in the platinum plot. In reaction chain 1, the waste buffer gout (after D1) is blended with glucose (the I2 droplet), diluting reaction chain 2. Concentration levels of glucose solvents in reaction chain 2 are lowered to half of their platinum levels (400, 200, 100, 50, 25, and 12.5 mg/dL). Similar effects may also be seen in chain 4 when the S3 waste buffering droplet is combined with the I7 glucose solution droplets—the sharp curve in the malignant Attack 2 calibration graph.

5.5.5 Drug Performance

Over the years, several high-profile situations have arisen that led to championships being dropped and the tournament being suspended. Many of these drug doping charges emerged after the test failed, and drug acceptance was revealed after that. In certain situations, doping is performed systematically on a massive scale, such as in East Germany and, more recently, Russia. But how efficient these exams are is an unsettled topic. What's the wrong positive rate? What's the false-negative rate? Is it possible to hack the organizations and tools utilized in drug monitoring? And are the quantitative rationale for these drug screenings problematic?

To handle general pro guidelines and legislation and to provide monitoring and training for athletes, the World Anti-doping Agency (WADA) was established in 1999. WADA provides the ISTI, which outlines, among other things, protocols for the performance of athlete communication for out-of-competing testing. WADA releases the ISTI. Practices for the athletes have begun. The examination may be easier to schedule and less invasive if POC diagnostic tools were available. Still, this type of use allows athletes and instructors to manipulate equipment or even use samples that are not part of the athlete-under-test. The ISTI also notes that "The certification organization for the specimen collection session concerned owns specimens from a participant." This might raise a problem of property and security because the drug testing method should only answer whether or not athletes utilized prohibited substances. The prospect of identifying and disclosing irrelevant information might enable the removal of whole specimens. The authenticity of sport and the ease, convenience, and confidentiality of athletes might be benefited from a reliable POC testing ground.

Change in features can be done via tampering, manufacture of falsified devices, or neglect by an end user. The hardware or software elements can be modified at all stages of a project round. A state involved in widespread doping fraud is likely to use its technical ability to change diagnostic equipment's functioning. Powerful vulnerability scanners will unavoidably demand security and trust approaches to such workloads.

5.5.6 DNA Fingerprinting Techniques

The use of DNA profiling techniques to detect persons in police prosecutions includes the implementation of DNA forensics. DNA collection is carried out first, then extraction, amplification, detection, and storage or destruction in the conventional forensic DNA flow is done. Each stage of DNA processing must be done in a pathology lab specifically equipped. Forensics laboratory regulation in the United States is fragmented, with just a few states which need forensic DNA accreditation. In addition, forensic DNA processing problems were discovered even with certification and government supervision, such as microbial spoilage and analytical neglect. DNA forensics development and adaptation of microfluid technology was driven by its ability to tackle the weaknesses inherent in standard laboratory flows and save time and expense, as shown in Figure 5.1.

Microfluidic devices have successfully executed cell lysis, DNA separation, extraction, polymerase chain reaction, and detection. However, all these procedures are not integrated into one mobile lab-on-a-chip device by the present state of the art. In this writing, practical use of this technology is limited. The microfluidic innovation is projected to reach maturity so that a crime scene device may be implemented with reliability and performance in mind. Automation and the reduced number of project phases in the examination decrease the probability of contamination or deliberate fouling. On the other hand, new technology might introduce new, unexpected attack avenues. The nascence of microfluidics in forensics with DNA offers the possibility to build security and confidence in critical applications, such as crime and government. In the fictional case shown in Table 5.1, Suspect A is excluded from the crime scene sample source. On the other hand, Suspect B matches the crime scene sample at 13 STRs. A calculation of Suspect B's genotype frequency, based upon the STR allele frequencies within Suspect B's ethnic group, reveals that the likelihood that a random member of this ethnic group has this profile is about 1 in 1.5 billion. It is essential to understand that this number is the probability of seeing this DNA profile if the crime scene evidence did not come from the suspect but another person.

5.6 Conclusion

The safety and confidentiality of CPMBs encompass new physical modalities and exciting new areas of application, comprising classic hardware security components, automation, and control system safety and more conventional information security methods. Due

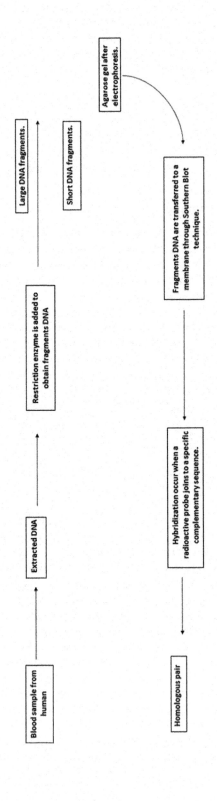

FIGURE 5.1
DNA fingerprinting technique.

TABLE 5.1

Example DNA Profiles Showing the STR Alleles for Each Sample and the Genotype Frequency of Suspect B for Each STR Locus

STR locus	Evidence sample	Suspect A	Suspect B	Suspect B's genotype frequency for each STR
D3S1358	15, 17	17, 17	15, 17	0.13
vWA	15, 16	18, 19	15, 16	0.22
FG	23, 27	21, 23	23, 27	0.31
D8S1179	12, 13	14, 15	12, 13	0.34
D21S11	28, 30	27, 30.2	28, 30	0.06
D18S51	12, 18	14, 18	12, 18	0.11
D5S818	13, 13	9, 12	13, 13	0.29
D13S317	12, 12	12, 12	12, 12	0.21
D7S820	10, 11	9, 10	10, 11	0.26
CSF1PO	8, 11	11, 12	8, 11	0.18
TOX	7, 8	8, 8	7, 8	0.30
THO1	9.3, 9.3	6, 9.3	9.3, 9.3	0.38
D16S539	9, 13	11, 12	9, 13	0.10

to the complex multidisciplinary structure of the system components, the challenges are complex, yet those new characteristics create innovative countermeasures. In addition, we argue that hardware security is required: the analysis of vulnerabilities resulting from the decision to design hardware and the creation of remedies in both equipment and software. It addresses various security problems that may arise on specified microfluidic platforms produced, marketed, or provided under open-source licensing. This article is by no means exhaustive; curious readers can instead consult up-to-date scholarly works to learn more about the capabilities of today's biochips. It gives examples of scenarios under various attack surface and assault results. It has several distinctive features that microfluidic bio-chips can be employed or exploited.

References

1. J. Tang, M. Ibrahim, K. Chakrabarty, R. Karri. Security trade-offs in microfluidic routing fabrics. In *IEEE International Conference on Computer Design*. New york. pp. (25–32). (2017).
2. K. Rosenfeld, E. Gavas, R. Karri. Sensor physical unclonable functions. In *IEEE International Symposium on Hardware-Oriented Security and Trust (HOST)*. London. pp. (112–117). (2010).
3. G. Zhang, C. Yan, X. Ji, T. Zhang, T. Zhang, W. Xu. DolphinAttack: Inaudible voice commands. In *Proceedings of the ACM SIGSAC Conference on Computer and Communications Security*. USA. pp. (103–117). (2017).
4. S. Bhunia, M.S. Hsiao, M. Banga, S. Narasimhan. Hardware Trojan attacks: Threat analysis and countermeasures. *Proc. IEEE*. Vol. 102(8). pp. (1229–1247). (2014).
5. D. Shahrjerdi, J. Rajendran, S. Garg, F. Koushanfar, R. Karri. *Proceedings of the 2014 IEEE/ACM International Conference on Computer-Aided Design*. (IEEE Press, Piscataway). pp. (170–174). (2014).
6. S.S. Ali, M. Ibrahim, J. Rajendran, O. Sinanoglu, K. Chakrabarty. Supply-chain security of digital microfluidic biochips. *Computer*. Vol. 49(8). pp (36–43). (2016).

7. L.R. Volpatti, A.K. Yetisen. The commercialisation of microfluidic devices. *Trends Biotechnol.* Vol. 32(7). pp. (347–350). (2014).

8. F. Su, K. Chakrabarty. High-level synthesis of digital microfluidic biochips. *ACM J. Emerg. Technol. Comput. Syst.* Vol. 3(4). pp. (1–10). (2008).

9. T.M. Squires, S.R. Quake. Microfluidics: Fluid physics at the nanoliter scale. *Rev. Mod. Phys.* Vol. 77(3). pp. (977–980). (2005).

10. D. Agrawal, B. Archambeault, J.R. Rao, P. Rohatgi. The EM side-channel(s). In B. S. Kaliski jr, C. k. Koc, Christof Paar(eds.) *International Workshop on Cryptographic Hardware and Embedded Systems* (Springer, Berlin). pp (29–45). (2002).

11. S. Checkoway, D. McCoy, B. Kantor, D. Anderson, H. Shacham, S. Savage, K. Koscher, A. Czeskis, F. Roesner, T. Kohno. Comprehensive experimental analyses of automotive attack surfaces. In *Proceedings of the USENIX Security Symposium.* (USENIX Association, San Francisco). pp. (77–92). (2011).

12. S.S. Ali, M. Ibrahim, O. Sinanoglu, K. Chakrabarty, R. Karri. Microfluidic encryption of on-chip biochemical assays. In *IEEE Biomedical Circuits and Systems Conference (BioCAS)* (IEEE, Piscataway). pp. (152–155). (2016).

13. A. Cardenas, S. Amin, B. Sinopoli, A. Giani, A. Perrig, S. Sastry. Challenges for securing cyber-physical systems. In *Proceedings of the Workshop on Future Directions in Cyber-physical Systems Security.* Berkeley. pp. (52–60). (2009).

14. R. Langner. Stuxnet: Dissecting a cyberwarfare weapon. *IEEE Sec. Prev.* Vol. 9(3). pp. (49–51). (2011).

15. N. Ferguson, B. Schneier, T. Kohno. *Cryptography Engineering: Design Principles and Practical Applications.* Wiley, Hoboken. (2011).

16. T. Xu, K. Chakrabarty. Functional testing of digital microfluidic biochips. In *IEEE International Test Conference* (IEEE, Piscataway). pp. (1–10). (2007).

17. R. Anderson, M. Kuhn. Tamper resistance—A cautionary note. *Proceedings of the Second USENIX Workshop on Electronic Commerce.* Vol. 2. pp. (1–11). (1996).

18. D. Grissom, K. O'Neal, B. Preciado, H. Patel, R. Doherty, N. Liao, P. Brisk. A digital microfluidic biochip synthesis framework. In *IEEE/IFIP 20th International Conference on VLSI and System-on-Chip (VLSI-SoC)* (IEEE, Piscataway). pp. (177–182). (2012).

19. D. Grissom, C. Curtis, S. Windh, C. Phung, N. Kumar, Z. Zimmerman, O. Kenneth, J. McDaniel, N. Liao, P. Brisk. An open-source compiler and PCB synthesis tool for digital microfluidic biochips. *Integer VLSI J.* Vol. 51. pp (169–193). (2015).

20. M. Rostami, F. Koushanfar, R. Karri. A primer on hardware security: Models, methods, and metrics. *Proc. IEEE.* Vol. 102(8). pp. (1283–1295). (2014).

21. H. Chen, S. Potluri, F. Koushanfar. BioChipWork: Reverse engineering of microfluidic biochips. In *IEEE International Conference on Computer Design (ICCD)* (IEEE, Piscataway). pp. (9–16). (2017).

22. M. Pollack, A. Shenderov, R. Fair. Electrowetting-based actuation of droplets for integrated microfluidics. *Lab Chip.* Vol. 2(2). pp. (96–101). (2002).

23. S.S. Ali, M. Ibrahim, O. Sinanoglu, K. Chakrabarty, R. Karri. Security assessment of cyber-physical, digital microfluidic biochips. *IEEE ACM Trans. Comput. Biol. Bioinform.* Vol. 13(3). pp. (445–458). (2016).

24. Preety, K.S. Kaswan, and J.S. Dhatterwal. "Fog or edge-based multimedia data computing and storage policies". In book entitled "Recent advances in Multimedia Computing System and Virtual Reality", Published by Routledge Taylor & Francis Group, December 2021. ISBN No. 9781032048239.

25. P. Sykes, S. Neoh, M. Brisco, E. Hughes, J. Condon, and A. Morley. Quantitation of targets for PCR by use of limiting dilution. *Biotechniques.* Vol. 13(3). pp. (444–449). (1992).

26. Preety, J.S. Dhatterwal, and K.S. Kaswan. "Securing big data using big data mining". In the book entitled "data-driven" published in Taylor Francis, CRC Press, 2021. ISBN No. 9781003199403.

27. M.A. Unger, H.-P. Chou, T. Thorsen, A. Scherer, S.R. Quake. Monolithic microfabricated valves and pumps by multilayer soft lithography. *Science.* Vol. 288(5463). pp. (113–116). (2000).

28. S. Dube, J. Qin, R. Ramakrishnan. Mathematical analysis of copy number variation in a DNA sample using digital PCR on a nanofluidic device. *PLOS ONE*. Vol. 3(8). pp. (76–86). (2008).

29. J.F. Huggett, S. Cowen, C.A. Foy. Considerations for digital PCR as an accurate molecular diagnostic tool. *Clin. Chem.* Vol. 61(1). pp. (79–88). (2015).

30. M. Zarrei, J.R. MacDonald, D. Merico, S.W. Scherer. A copy number variation map of the human genome. *Nat. Rev. Genet.* Vol. 16(3). pp. (172–180). (2015).

31. A.J. Lafrate, L. Feuk, M.N. Rivera, M.L. Listewnik, P.K. Donahoe, Y. Qi, S.W. Scherer, C. Lee. Detection of large-scale variation in the human genome. *Nat. Genet.* Vol. 36(9). pp. (949–960). (2004).

32. G. Wang, D. Teng, Y.-T. Lai, Y.-W. Lu, Y. Ho, C.-Y. Lee. Field-programmable lab-on-a-chip based on microelectrode dot array architecture. *IET Nanobiotechnol.* Vol. 8(3). pp. (163–171). (2013).

33. M. Walker, M. Chi, N. Navin, R. Lucito, J. Healy, J. Hicks, K. Ye, A. Reiner, T.C. Gilliam, B. Trask, N. Patterson, A. Zetterberg, M. Wigler. Large-scale copy number polymorphism in the human genome. *Science.* Vol. 305(5683). pp. (525–528). (2004).

34. J.F. Huggett, C.A. Foy, V. Benes, K. Emslie, J.A. Garson, R. Haynes, J. Hellemans, M. Kubista, R.D. Mueller, T. Nolan. The digital MIQE guidelines: Minimum information for quantitative digital PCR experiments publication. *Clin. Chem.* Vol. 59(6). pp. (892–902). (2013).

35. K.S. Kaswan, L. Gaur, J.S. Dhatterwal, K. Rajesh. "'AI-based natural language processing for generation meaningful information Electronic Health Record (EHR) data' published in Taylor Francis, CRC Press for 'advanced artificial intelligence techniques and its applications in bioinformatics'". (August 2021). ISBN No. 9781003126164J.

36. M.I. Tang, K. Chakrabarty, R. Karri. Securing digital microfluidic biochips by randomising checkpoints. In *Proceedings of the IEEE International Test Conference (ITC)* (IEEE, Piscataway). pp. (1–8). (2016).

37. G. Chen, Y. Lin, J. Wang. Monitoring environmental pollutants by microchip capillary electrophoresis with electrochemical detection. *Talanta.* Vol. 68(3). pp. (497–503). (2006).

38. J.C. Jokerst, J.M. Emory, C.S. Henry. Advances in microfluidics for environmental analysis. *Analyst.* Vol. 137(1). pp. (24–34). (2012).

39. S. Neethirajan, I. Kobayashi, M. Nakajima, D. Wu, S. Nandagopal, F. Lin. Microfluidics for food, agriculture and biosystems industries. *Lab Chip.* Vol. 11(9). pp. (1574–1586). (2011).

40. J. Wang. Microchip devices for detecting terrorist weapons. *Anal. Chim. Acta.* Vol. 507(1). pp. (3–10). (2004).

6

Cyber Physical Systems in Supply Chain Management

Rajbala, Pawan Kumar Singh, and Avadhesh Kumar

CONTENTS

DOI: 10.1201/9781003220664-6

6.1 Introduction

Individually tailored commodity requirements contribute to the difficulty of supply chain preparation and management (SCs). Digitizing business processes is a crucial development and a necessary prerequisite to achieving optimum flexibility and performance. In particular, cyber physical systems (CPSs) are increasingly essential for the digital transformation of SCs. Industry 4.0 principles and innovations (I4.0) will become a central facilitator. Implementing such technology also ensures that an enterprise may have disruptive consequences so that management supports change processes. In this sense, the management challenge to deal with improvements has grown increasingly, along with the construction of digital systems' features and functionalities. CPS can provide a wide range of technological equipment and specifications for various applications. Support for the design and development of CPSs in SCs is required [2].

6.2 Digitalization Transformation of Industry 4.0 of Supply Chain Management

Modern electronic communication systems are used for the present synchronization of networked SCs. The hypernym "Industry 4.0" sums up this digitalization of the industry in Germany. Even though the term "Industry 4.0" has been in existence for many years, no uniform definition exists. Bischoff gives a systematic description suited to the extent of the paper. He describes I4.0 by connecting the physical and the online environment as developing processes for output and increases in productivity. This connection is provided by self-controlled CPSs, which allow horizontal and vertical implementation to produce the goods or services of the SC efficiently, decentralized and versatile. Digital transformation is considered an essential criterion for obtaining I4.0. Thus, businesses in an SC create digitization plans to coordinate their activities and digital transformation campaigns. Enhancing an organization's success and reach using new technology is digital transformation. The following description applies to and can be used for the digital transformation of operations. Using new technology to sustain or increase productivity means that digital innovation combines improvements in business processes [3].

6.3 What Is a Cyber Physical System?

The above description explained that a CPS plays a significant role in I4.0, forming the foundation and leading technologies. The part of the CPS is to connect the digital and physical environments. They can continue supporting and performing tasks independently by planning and controlling the SC. Lee is one of several and most commonly used CPS interpretations: Cyber-physical structures are collaborations of complex physical processing. Incorporated computer networks usually supervise biological phenomena with feedback mechanisms when physical procedures impact calculations and vice-versa. This characterization is not enough since special techniques are also explained as CPSs in this interpretation. Broy also explains that they can communicate and thus collaborate as a core attribute of CPSs. Nevertheless, the news does not have available communication capacity in the medium, as has been popular in autonomous vehicles for computational mathematics systems for years, in the form of multiple routes. Instead, the many CPS renewal builds relationships via open and global communication technologies, especially the internet. Open and connected systems that connect with metric quantities—virtual reality with the actual operation world is beginning to emerge. As a result, CPSs can start taking over management and execution tasks in SCs while also reacting effectively through collaborative efforts in real time. Apart from the descriptions, the dictionary diagram is valid and applied in the context of planning and scheduling in SC: Socio-technical structures are a way of interacting with the various systems; pushing to improve relations between internal and foreign clients in SC is possible to lessen interface issues and to mitigate maximum interoperability issues.

6.4 CPS Innovations

Several technological elements and individual functions can challenge the development of the CPS. These frequently rely on the scope and intent of the CPS. Not all innovations should be sensibly used in preparation and control depending on the weather of a process. Various potential solutions can also be developed according to the scheduling and control mechanism specifications in the design of a CPS.

6.4.1 Features of CPSs

CPSs have procedure features and technological features. Given that CPSs are mostly embedded with current organizational structures and are closely linked with the flow of materials and knowledge, process features play a vital role throughout the design phase. The categorization and systematizations presented are the basis of these characteristics. Method planning and monitoring in SCs is distinguished by decisions that can be found and followed in various manners. The development of CPSs means that the separation of duty between humans and the system must be established. The first feature to be listed is the source of protection for decision-making. The following terms may be rendered dependent on humans, decentralized and machines.

Coordination is needed where there are interrelationships and interactions among decision-making components of a structure, including functional levels and procedures, which can be taken as a consequence of the division of work. It differentiates between hierarchy and hierarchical collaboration and intermediary events. The maximum level of self-regulation can be reached in hierarchical teamwork; in its work on physical sensors, the individual components of a structure organize each other in decision-making and develop this feature by separating vertically, together, geographically, between three modes of synchronization.

Using technology, a CPS can execute different functions. This will help the management and forecasting activities in SCs incredibly. Functional criteria characterizing the behavioral and additional features of a device are decided by developing CPSs. Based on the phylogenetic box taxonomy according to the distinctive skill of a CPS, time consumption, data management, data gathering, data capturing, data provision, acting, decision-making, networking are provided with the expresses of the stereotypical "function."

Environmental conditions can adversely affect the activity of a CPS. These can be largely subjective depending upon the type of implementation and can significantly hinder the incorporation or function of a CPS. Also, under dynamic environmental circumstances, it must be assured that CPS works efficiently. External conditions include issues that can affect the part of the system, especially in a positive relationship. The detrimental variables in this document were pointed to as the main characteristics of the negative effect on a CPS in the system. Because CPS represents a mixture of digital and physical features, no destructive force, material flux factor, or knowledge flux factor can be expressed.

The adverse impacts of a company's supply chain are directly associated with the objects. They are the simplest CPS methods. Many other physical artifacts, such as materials, structures, means of transport, manufacturing facilities, or logistics elements, are found in SC processes and may thus be considered procedure CPS's features. The scope is focused on material items in the SC to minimize the available subjects. The main concern of such artifacts is a CPS in terms of manufacturing and logistical capabilities. In addition, both fixed and mobile artifacts can be used. The following expressions are given to the specific "physical entity": apparatus, work assistance, location, substance, and person.

The technical elements in the CPS are closely connected to the method features. These technological advancement features allow individual functions to be performed with the help in SC's planning and control activities. There is no uniform description of the CPS, and even the technical parts or features of such structures are not consistently specified. Besides the Lee definition, the documentation also proposes several other CPS meanings. Roy gives an outline of his work on different CPS concepts. A CPS literature review is also performed to classify the most relevant technical elements of a CPS, based on further descriptions of a CPS. There are additional phylogenetic features of the box. The essential technical aspects of a CPS can be determined based on the meanings. A tangible entity situated in the process setting constitutes the foundation of each CPS. These are complemented by microcomputers, sensors, and drives and developed into embedded systems. On that basis, it becomes a CPS through communication systems, such as the internet, clouds, other subjects and techniques. Below are the technical features which reflect the characteristics of the morphological box. Additional literature research shall be used to define the quality.

The physical thing must be aware of its atmosphere and condition to equip an object with information. Sensor technology provides the opportunity to monitor and measure real-world parameters and translate them to electronic pulses. For I4.0, sensors are a vital precondition for and above technological improvements. Sensors, embedded and intelligent sensors, and wireless sensor networks may possess essential technical attributes: identifying sensors.

Actuators allow interaction with the world and affect it. They will enable the reaction of material things. Closing controller actuators are an essential feature for the automation of operations. Their skills can differentiate them. Responses, physical activity, and user engagement are included in the anatomical boxes of the following words.

The cognitive processes are another technological aspect of a CPS and hence a feature of the morphologic box (see Figure 6.2). The measures required to manipulate a state variable are generally decided in the construction of knowledge (VDI 2004). Their systematizations discuss the place of knowledge discovery. Three types of data processing can then be established: subcontracted, together, and incorporated.

Different humans and machines may interact with a CPS using a variety of physical things. One term is the relationship with the complex organism. When functions and procedures are increasingly complicated, appropriate technology must assist people.

However, virtual interfaces exist between other physical devices and information systems because only connectivity and communication allow efficient CPS potential optimization. The development of such an interface may be based on the following characteristics: contact between the person and the computer and interaction between the machines. They are linked in a network to allow physical structures, persons, or systems to communicate and share data. The internet is a critical network in CPS communications. However, contacts do not need to be carried out across a worldwide open network. Therefore, a wired, cellular, mobile network would depend on the form and type of network link.

The CPS literature review indicates that clouds are often referred to in CPS contexts. A shared interface allows local physical items to be connected globally with others. Cloud computing facilitates connectivity between those artifacts through companies by offering them technology, including applications, data storage, and services via data center links like the internet open to users of these IT resources. Various operator models are expressed as cloud-based: private, public, and mixed cloud. There are no different models available.

6.5 Designing and Importance of Cyber Physical Systems

In the steel corporation's manufacturing and supply sector, the morphology box is used as a creative method in designing CPSs for two uses. In seminars with experts from the respective field, the identification of the program director was conducted. A total of eight CPS-based scheme materials for different procedures have been formed, two of which are discussed below.

The first instance of use is a checking of goods receipt in which the CPS will allow us to identify deliverable plaques. The plates are carried by rail from Rotterdam harbor to a company's facilities in Germany. The commodity parts cannot currently be accurately classified, and they could be altered on the transportation path between seller and consumer. Therefore, the platforms cannot be controlled, and the storage cannot be accomplished. The labels should be fitted with an RFID tag shortly. The number of the material part is stored, making the pieces unique. The rail is passed through an RFID gate in the receiving zone to read out the RFID tags and return identity information. After retroactivity of the data, the locomotive engineer is provided with the lead through a data warehouse with ordered items to save the slabs. It is primarily managed by distributed decision-making and vertical collaboration. The relevant data from the warehouse is showcased via a projector to the forklift driver and made accessible to the control scheme. Communications between the processes are primarily interconnected, and central, inner database management systems provide, store, and analyze the information. In the second instance, the method involves producing steel coils and transporting them internally through both processing stages. The process is highly resistive and results in a high cost of cooperation and long time frames. In the future, a remote system network will link multiple manufacturing facilities, buffer areas, manufactured coils, troop carriers, and cranes. These assets can clearly define, record, shop, and inform process flow data. This data is conveyed to operative workers via smartphones on the physical subjects decentralized. The manufacturing unit upstream is coordinated with the downstream manufacture planning of the machine, which contains information such as maximum throughput use, faulty work, or manufacturing programmer. Such documentation is hosted in the server, analyzed through technical working, and made accessible to participants. After manufacturing the coil at the first stage of processing, the structure begins its transitional storage by interacting with the independent transport and crane system. Similarly, the downstream manufacturing unit is responsible for the recovery of the required coil—this whole system is part dependent, and operation and lateral coordination are based on machinery. Critical essential connectivity usually occurs via a cell network, which can often contribute to issues with data transmission quality in production halls. In addition, excellent physical and mechanical stress is placed on steel belt manufacturing.

6.6 What Is Supply Chain Management (SCM) 4.0?

"Supply Chain 4.0" means physical, technical assimilation of network-wide systems that enable increased production, organization, and financial performance, typified by autonomous actions that are location-independent, prevalent integration, numerous technological services, and by their ability to respond in a context to customer demands and requirements. Specifically, this word coined the industrial development 4.0 to signify the industrialization and the absorption of intelligent technology in suppliers and customers as shown in Figure 6.1. These systems promote the growth and functioning of industries and government, highlighting global network exchange and safe ownership of templates and information [5]. For firms, supply chains are necessary to meet critical logistical and operations.

Supply Chains 4.0 are more like a kind of tactical capability and may perhaps have a more extended mission and life-critical implications. Companies are now based on industrial information platform networks for digital operations every day. This dependence creates cyber insecurity as complicated cyber and physical hacker situations breach those systems. The advancement and security of cyberattacks is a research field internationally involved.

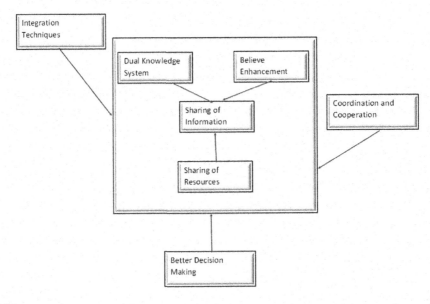

FIGURE 6.1
Framework of SCM 4.0.

The capacity of both the government and the non-national state continues to expand and increase. Around the same time, Supply Chains 4.0, integrated and technologically enabled [6], are getting ever slimmer. A significant field of study is created by the connection among digital technologies with the ever-more militarized cyber area. Depending on computing and information brains, the operation and accessibility of supply chain architectures are changing. The disposition of a military supply chain by leveraging a weakness will lead to economic implications and threaten human life [7]. As security materials are widely procured, the defensive surface for malicious actors is widening internationally, resulting in the possible increase in the impact of a cyber assault on supply chain networks 4.0 [4].

Even more than before, broadband and action potential are critical to the global Supply Chain 4.0. This reliance undermines their safety. Changes that have taken place in functioning, architecture, and network administration reaffirm the need and future importance of global value chains. The supply chain risk is classified as the unpredictable likelihood of altering the macro or the supply chain systems, which influences any element of the supply chain operations (including IT and OST). Risk management, cyber threat forecasting, and risk identification management frameworks to avoid or lessen its impact [8] assist in showing a supply chain cyber risk evaluation 4.0. There are two analytical forms of vulnerabilities inside the supply chain: interruptions and service [9]. The risk of disruptions is not reduced by any natural disaster, for instance, storms or floods. Operational risk, such as cyber assaults, deals with failed production and consumption procedures in manufacturing or delivering finished goods.

Project management of security risk, agile technology, and risk control will help standardize the supply chain procedures 4.0 and mitigate mission success [10]. It also enables technological innovations to be implemented safely, blockchain, the Internet of things (IoT), artificial intelligence (AI), the cyber physical systems, etc., for example, the Internet of things (CPS) [11] are providing companies with automatic and reliable resources, increasing efficiency and supply chain flexibility 4.0. Models of mission assurance will alter the response of the military and defense organizations. Themed technologies such as crownjoy research, the methodology for high availability, online semantic models, operation orchestration structures, and upgrading obsolete or deteriorated systems and procedures can increase organizational task security. The technology world impacts how military operations and competition in this area will provide challenges and opportunities for the efficiency of operations in the military supply chain. As a result, defense companies increasingly need to assess how emerging technology will influence their supply chains to implement or counteract them as required.

This caption focuses on the singularity of how the armed forces function, focusing on Supply Chain 4.0 and their stable assimilation with technological advances. The stress is less on their information systems than on data they contain and customer data. They are supporting elements. It includes an extensive literature review on the link among cyber security, security, Supply Chain 4.0, and semantic modeling. In this regard, this chapter describes:

- The proper business goals of the supply chain management army in light of the developing world of interconnected forces;
- An analysis of the current condition of the chain management military and defense networks;
- Logistical support;
- Various ways to measure the effect of emerging developments on the military supply chain.

It summarizes evolving Supply Chain 4.0 principles, connects them to emerging business and the army demands, and outlines the need for semantical modeling to consider the issues and problems in this space. Eventually, this work describes the core developments and technology interactions with growing technical interconnectedness in this field.

6.7 Concepts of Supply Chain 4.0

The history of production processes and their predecessor Supply Chain 4.0 is discussed in this section. The importance of the 4.0 supply chain for military and industrial applications has also been emphasized in this segment.

6.7.1 Logistics in Supply Chain Management

Shows assimilation by the physical supply network between customers, distributors, tools, manufacturing plants, and computer science [32].

Supply Chain 4.0 creates an alteration that rethinks the conception of the electronic supply chain of businesses. Because of consumer demands for speed, efficiency, and openness, several approaches have been developed to upgrade current procedures. In addition to adaptation, supply chains can improve operational productivity dramatically and take advantage of the benefits of new market models of the digital supply chain. Supply chains must be quicker, open, precise, and agile to produce their advantages.

Various analytical approaches were explored, anecdotally the first step in many agreed risk management systems to determine cyber risks from the supply chain. The writers in the article say that augmented intelligence from cyber threats would help Industry 4.0 solutions further identify and consider cyber hazards. One of the most significant issues created by earlier research was the data sources or actual data that allow obtaining credible conclusions from supply chain network cyber risk occurrences. The supply chain also influences the information management of technologically advanced products. Usefulness and supply chain management steps inevitably lead to device and performance measurement that results in a uniform network that raises vulnerabilities and decreases the challenges to an intruder. The foundation for these problems contributed to unique field-specific buildings in the cyber supply chain risk management.

This cyber risk inventory control covers activities for assessing and mitigating later operations (planning, programming, manufacturing, integrating, and implementation included). The IT channels' equipment/software structures comprise supply chains. Efficient and realistic supply chain control provides network and corporate connections throughout the entire chain. The sophistication of distribution networks raises uncertainties that spread across the whole network and possibly affect company operations. The supply chain models have to provide concerns for retailers', distributors', and clients' volatility to politically mitigate the adverse effects of instability in the supply chain. An institution's capacity to maintain the efficiency of the supply chain depends on commodity consistency, pace of business execution, and agility changes depending on the need of the consumer.

Improving the level of connecting nodes in each supply chain is the fundamental concept of supply chain management, thereby reducing service costs or the time required to modify the environmental stimuli and adapt. Coordination is essential for lowering product stocks, removing limitations, and achieving high standards. Enhanced awareness of the supply chain enables supply chain operators with prompt and reliable information. This increased availability to the system requires understanding and dynamic response and improves the supply chain value stream, allowing waste to be reduced and efficacy to be improved. Therefore, an integrated design of lean business models and skills is necessary to efficiently and effectively operate global supply chains.

6.7.2 Supply Chain 4.0 in Defense Combat Action

Supply Chain 4.0 enables drug interventions, operations, and strategies. Combat actions cannot be maintained without practical and effective logistics operations. Military suppliers have various effects on civilian counterparts and start operating with a distinct final state to consider. Army needs to include confrontation preparedness, the flexibility of the supply chain, diversity of items, and unpredictable requests. A conventional supply chain is the network of goods and services, processes, and organizations that provide value to the end users of them. Conventional forces are the great candidates between nations, distributing and transportation systems that build a world network as a whole, as shown in Figure 6.2. The supply chains include "material flows, products and knowledge flows, interconnected by various direct and indirect facilitators, namely partnerships, procedures, operations and coordinated information structures inside and between organizations." Their technology base includes delivery systems, channels and connectivity networks, and physical delivery networks. Supply chains are usually subdivided into three phases: sourcing, manufacturing, and delivery [13].

Supply chains are precisely the definition of logistics but are distinct. As part of the continuing and interconnected operation, supply chains combine procurement, processing, and delivery. The logistic "point of origin to the point of usage" is seen as just part of the convergence of logistic supply chains into all primary market operations [14]. This includes process integration and coordination, an electronic data exchange, and sustainable strategic relationships. Production methods have evolved scaled versatility from wide stocks with low advance notice and deferred commodity setups that deliver optimum value for capital and a high degree of versatility when combined with demand management practices.

6.7.3 Significance of Supply Chain 4.0

Supply Chain 4.0 is the convergence of production and communication technology to enhance the development of conventional supply chain networks through independent

FIGURE 6.2
Supply Chain 4.0 in defense combat action.

intervention, prevalent integration, a variety of integrated resources, and the ability of traditional supply chain systems to respond in a sense to consumer preferences and needs. The depth of the supply chain in Industry 4.0, as seen in Figure 6.1, demonstrates the difference between expected noncombatant outcomes [15]. The assaults on armed forces' supply chains often significantly target these specific standards to affect military activities. The results of a 4.0 supply chain failure vary according to the meaning and quality of a supply chain. Whereas there are many advertising distribution channels for the supply of profit-enabling goods [16], defensive system product lines can affect a force's management and strategic capacity, which directly influences a person's existence and the achievement of a quest. In contrast to many businesses, military devices are less geared to maximizing profits and are increasingly security-oriented, leading to immediate and military conflicting interests [17]. Supply chains, as well as strategic thinking, are closely connected. Modifications in global value chains are flowing into the military-strategic plans. The enhanced capability to deploy troops, prevent complications, and deliver more competent. Interoperable logistical forces are linked to the objective of optimization of the enterprise structure of the logistics data system. The organization, monitoring, and final execution of strategic distribution and logistics features demand highly integrated, secure logistics information technology. For example, the Royal Australian Air Force document describing their commitment to procurement does not discuss the safeguarding of the security supply chain and does not discuss computer security.

In the strategic edge of companies, this global security supply chain is an interactive mechanism that allows enterprise systems to turn capital into goods and services. Increased computing supply chain management by greater mobility, process automation, supply chain virtualization and digitization and responsiveness to the markets was achieved in the shift to a knowledge economy that not on commodities but knowledge. The knowledge exchange across the supply chain inevitably contributes to a virtualization of the supply chain paradigm that logically leads to process convergence. Customers, manufacturers, dealers, and product designers are integrated into the supply chain through standard processes and shared knowledge. The idea of an expanded company depends on more incredible energy and exchange of knowledge. The advantage is that the corporate value chain is extended to and through other markets and organizations. The convergence of network players and nodes into a single infrastructure creates a different degree of competitiveness, which will build on the networking participants' advantages and skills to act more effectively on business demands in a competitive way.

While the military continues to imitate manufacturing worldwide to reduce stocks and improve performance, several steps have been taken in roles and procedures. The authors argue that there continues to be more room for progress in the alignment of the supply chain across both DoD and its providers—processes that influence the net cost and efficiency of the supply chain collectively. This is a significant objective that includes several vendors and businesses. These initiatives have been strengthened in several ways. The CORE project aims to realize effective, quick, and accurate supply chain solutions. CORE has several application demonstrations to illustrate how symbolic operational scenarios can accomplish a safe global supply chain. The WEF has established the World Economic Forum, a draft of alliances, policies, strategies, and technologies for resilient supply chains [18].

Various evolving nearest cluster centers and conceptual models aim for market usability, military supply chains, or emerging industries like Industry 4.0 to be defined and semantically formed. Business Process Model and Notation (BPMN) 2.0 is the best-known method for amending business operations. Numerous BPMN extensions expand the notification formally established to include new concepts, including the cloud network [19] and the use of IoT platforms [20]. Processes to incorporate BPMN into combat actions and procedures are increasing, though are not accessible at the time of this writing for peer review and analysis. The Open Group Architectural Framework (TOGAF) [22] and the Department of Defense Architecture Framework are alternatives to BPMN (DoDAF). These degrees of complexity are higher than BPMN but generally agreed upon. During Industry 4.0, there is no detailed updating phenomenology capturing this field, given the exponential growth and rapid release schedules currently experienced [21]. Instead, one factor is based on most formal methodologies and conceptual frameworks in this field.

6.7.4 Dependencies on Supply Chain 4.0

Critical criteria drivers and considerable developments in the expansion of the industry supply chain, a harsh fundamental reality impacts the production and creation of defense capacity. The US Defense Department (DoD) has estimated that about 70% of electronics products are redundant, replaced, or out of demand in several procurement phases before the finished system is deployed. It's not only for system upgrades and overhauls, which also need the purchase of new parts.

Despite improved exposure across the interconnected supply chain, the clarity remains minimal for several reasons. Commercial drivers may, in some situations, impede the exchange of knowledge. Even though records reside within the supply chain, proprietary data security can hide details and minimize assurance. The size of global supply chains may, in some situations, operate toward the control time, such as Australia, where the volume of transactions is not necessary for international suppliers to exchange knowledge or work together. Computing, IT, and networking facilities are also used in supply chains. Governments worldwide have considered and implemented prohibitions on Huawei technology from governance and international forums regarding supply chain contamination and other associated flaws in purchasing IT products. This strategy was found both too large and ultimately futile because it would not resolve all kinds of attacks from the supply chain.

6.7.5 Challenges of Modern Procurement Policy in the Defense Sector

In defense organizations, a lack of current procurement policy describes business procurement. It thus makes it hard for insights from defense experience and the current market to be translated into a managerial paradigm application [23]. Cyber susceptibility can occur with supply chains as sophisticated systems (SoSs). Corporate consequences are mainly when supply chain components rely on the information from the Internet of things. Incorporating broader military environments of subnets, technologies and facilities of the supply chain create problems for their safe use—prospective strategies for mitigating device integration risk. The development of company architecture (EA) approaches [24] is part of dynamic supply chain processes. EA deployments' standard requirements include independence for policy intervention reductions, a proposed policy perspective if appropriate, and the ability to respond quickly to changes. In the past, support and assistance were an alternative to secure complicated EA, but they do not have a comprehensive solution [25].

In developed market economies, IoT convergence is a rising development, and it is anticipated that these technology companies will join military supply chains through the period. Thus, the security chain systems and networks can undermine the figure within IoT devices. The insecurity of IoT devices within supply chains is influenced by many factors, including limiting specifications and restrictions on resources, cloud storage and processing consequences, extensive data protection, and reduction in prices [26–28]. Though there are obfuscation techniques for these potential threats, the efficiency, financing, interoperability, required resources and flexibility, sophistication, and viability of the comprehensive method will cost them [29–31]. The complexity of military supply chain networks and technology poses considerable threats to their safety. Therefore, from the highest development of specific networks and modules to the global environment, protection in the supply chain must be established so that we don't ignore any loophole that could have disastrous implications for the militancy.

6.8 Modern Integration Systems of Supply Chain 4.0

SoS-based Supply Chains 4.0 operate and never exist as individual organizations. To be functional, these programs can then be combined, or the weak link will still be attacked. Knowledge sharing, teamwork, confidence, desire to work together, collaboration, and shared business interests are the most shared factors influencing the convergence of the supply chains [33]. The changing defense environment brings a wide range of variables impacting defense supply chain service, insecurity, and responsiveness. Interconnection and synchronization of networks, partners and outsourcing, life cycle components, threatening actors, and defense company problems are all factors to be considered.

6.8.1 Connection of Supply Chain 4.0

Military supply chains never operate as single entities but often involve merging several SoSs. The design of this device combination turns the management of these complex supply chains into a company's problem. External applications have their features, compatibility considerations, and accessibility requirements of both public and private organizations, which may pose challenges for integration. The SoS, which comprises a comprehensive Security Supply Chain 4.0, has no consistent frontier with the other information systems, with which it is unavoidably linked. Security supply chains cannot be regarded as sole bodies because they are mostly related to networks to notify, instead of shape, the supply chain.

The definition of the potential business environment affects the efforts and course of capacity growth. When assessing future power capacity and architecture and contemplating existing and future priorities, there is a great need to recognize the current operational climate. The military technology process information uses the term "extended attack flat" to characterize a greater area of the intrusion triggered by the mixture of networks, persons, and risks of growing amounts and variety. This chapter concludes:

> These programs have the greatest weakness to facilitate their usage and use: performance. Natural monopolies and reliability in the management of businesses are greatly appreciated, whether cost-cutting or the demand. As these networks face a surge of automation that drives performance, threat actors may be more easily attacked. Stated: hacking effectiveness is simple.

The working climate for tomorrow is also changing from emerging technology. Artificial intelligence convergence (AI) and machine education systems enhance the success, with emerging innovations being generated, transformed into Supply Chain 4.0 by drones and robots, together with the ubiquity and effectiveness of the Internet of things (IoT). The world's supply chains will benefit from any such advancement forward. This is defined as "the industry's networking contribution to developing the natural foundation for AI implementation and scaling that enhance the human constituents of global supply chains that are strongly organized." Such techniques pose their threats, making assailants more vectors and possibly permitting hybrid cyber and movie offensive actions within a military setting, AI services supporting attack tracking and speeding, and data armament. Military weapons are mostly purchased from suppliers of diverse and non-transparent supply chains, and parts are transited via the SoS before delivery to their ultimate destination. Accountability for the accuracy and credibility of each piece of goods will also be hard to assess. Therefore, electronic system shipping can be undermined before reaching the military Supply Chain 4.0. The incorporation of Supply Chains 4.0 is intended to allow a mutually advantageous ecosystem capable of capturing the convergence of enterprise applications between intra- and inter-entrepreneurs to maximize the general business operation of the organization.

6.8.1 Distribution of Supply Chain 4.0

A 4.0-enabled integration supply chain combines multiple silos to shape the whole supply chain or an end-to-end vision. The standard procedures in chain supply risk management have been observed for the combination of the operational technologies (OT) that manage systems for distribution and manufacturing electricity generation and navigation processes for production and control, with IT that administer software and connections, as well as the supply chain that assists suppliers, products. Integrating the economic and market advantages realized with greater productivity requires the distribution network. There is a benefit in world supply networks. Transparency from end to finish, from manufacturing to delivery convergence and bridge cooperation, improves the agility, stability, and slow-moving stock that is not achieved in massive supply chains.

The common operating picture (COP) is one step toward strengthening the supply chain, presenting the actual status of logistical processes with a single integration stage. In military programs, assignments, and procedures, COPs are widely used. However, a COP highlights how well our future technology may be integrated and automated by their designers. The degree of convergence into a fully integrated supply chain worldwide goes well beyond the notion of a COP. In short, military networks are not intended to deal with the scale and communication that potential supply chains will provide, which is usually used to track and control operations. The correlation between material and non-physical objects develops various limitations which, without instruments from the 21st century, are extraordinarily complicated to comprehend and handle. This will demonstrate the complex design of the applications since the risk is managed using Microsoft Excel tablets and PowerPoint. For example, the Joint Strike Fighter Program allows operative military forces to pay for success rather than replacement parts. Instead of the nation's airplane, Lockheed Martin has this multinational supply chain administration. This directly affects the independence of aircraft capabilities and the potential to support aircraft in a crisis involving the powers of the supply chain.

6.8.2 Supply Chain Partners and Externalization 4.0

Supply chains are not made up of a single company but rely on foreign retailers and sellers to buy their goods. Supply chains are made up of SoSs that never function as individuals.

Those systems must ultimately be implemented to ensure efficient and viable supply chain management methods. The most common factors impacting supply chain integration are knowledge exchange, teamwork, confidence, readiness to work together, collaboration, and shared business goals [34]. The armed force supply chains have four major risk categories; partner's risk, hidden risk disclosures, interest risk dispute, and risk control. They can be manifested by the lack of cooperation between organizations concerning cyber security requirements, influencing civilian allies, contradicting their expectations of the need to secure measures, and subcontracting the scattering of confidence.

Programmable interfaces for applications (APIs) that support program creation and use are typically available for app developers beyond the product company to customize a program for their individual needs. This speaks volumes about both the platform's lack of API compatibility regulation. The disconnect between strategic and trade supply chains can have different potential for weakness due to the mismatch of priorities with organizational results and profits, respectively [1]. A military violation of protection will have severe implications for civilians' lives and national sovereignty. Conversely, the events are expected to become more economically central in business activity. This is shown by a data breach of the Australian Security Force in 2016, in which assailants reached their targets by exploiting one of the ADF vendor vulnerabilities. The number and positions of the many manufacturers needed to manufacture a product may also threaten military supply chains. About 70% of foreign exchange contains semi-finished goods undergoing final remodeling. In the world economy and global supply chains, parts from various countries and suppliers may also be the most accessible goods. For example, a 600-meter Dell Inspiron Notebook includes components made in more than ten countries, including the Philippines, Malaysia. And the F-35 Strike Fighter, used by countries such as the United States, Europe, and Australia, includes components from all over the world and China.

6.8.3 Vulnerability of Supply Chain 4.0

The cyberspace world is rapidly changing, with frequent advances and vulnerabilities. The Common Vulnerabilities and Exposes database reported 3,297 weaknesses in 2018 alone. In both numbers and complexities, cyber threats are year on year increasingly advanced and have a range of consequences for people, companies, and government. The primary goal is a cyberattack vector relevant to security supply chain networks. It applies to cybercriminals that compromise enabled objectives to reach the enterprise with more significant benefit. An intruder will, in this situation, breach a military infrastructure by one of its supply chains associates, which have less established network defense.

Protection by anonymity is the best security practice from the historical viewpoint. Examples of protection by obscurity involve modification of default property parameters or the allocation to facilities of odd ports. The technology is often applied to applications unfamiliar to people like SCADA systems. Although dark safety can be a defensive strategy toward opponents, it is not adequate for a plan. Cybercriminals do not necessarily need to recognize a structure in its entirety to target it and achieve their objectives.

The number, efficiency, and scope of cyber-offensive operations have increased over the last decade. Advanced persistent threats (APTs), problem organizations, and crime syndicates are highly diverse and spend time and money on cyberattacks. It cannot be denied the demonstrated success of cyberattacks for military purposes. Given this efficiency, the supply chain is a possible objective, a central element of the new military: coordinated efforts in supply chain management, network security, government and commercial networks, and perhaps a large attack surface. Cyber security is similar to asymmetric combat, with the malicious players only having to rely on a safe system's weakest node.

Cyberattacks in trade areas often produce some kind of numerical result, either directly or indirectly, like using industrial spying or encryption. The 2017 attacks by Winery emphasize how financial insurgencies are motivated. The assault was thwarted by encryption to restrict access to data that could only be recovered after a monetary ransom was paid. The car industry had recently substantially displayed another example of hacking manual control systems with multiple integrated and perhaps internet-connected computer controls. This introduces additional cyberattack mechanisms in physical networks with no oversight or input into their function. There are no long-term effects on vehicles that integrate major ITs.

Such hidden cyberattack interfaces are also present in the case of drones, robotics, and other remote-controlled devices. In the future factory, cameras, drives, and autonomous robots are generally controlled by complete military activities. A cyber assault on transportation facilities could hypothetically affect activities, cause diversion of vital machinery on the way, movement of machinery, compromise routes or disturb device activities, and halt all flow of logistics.

Since essential facets of the potential supply chain are anticipated to be greatly simplified and cyberattackable, it is assumed that the safety perimeters have been broken for many military structures. If data is damaged by a threat or a spontaneous mistake, supportive logistics should be resilient and secure enough to ensure that data manipulation is continued to be adequately supported. The capacity of an opposing party to control or interrupt production lines are both of operational and supply chain financial capabilities excellence paths if a physical and cyberattack is orchestrated.

6.9 Supply Chain 4.0 Risk Management

This discusses the risk assessment process and the application to the 4.0 supply chain and tests its usefulness in a military supply chain situation.

6.9.1 Analysis of Risk

Structures for risk analysis help companies identify their possible vulnerabilities and evaluate how those vulnerabilities could impact the company and its processes. Analytical approaches determine whether the network security priority of the enterprise is to avoid an attack in a first case, to react when an intrusion occurs efficiently, or to both. The evolving scope of risk reduction has changed from the security of assets to business processes or the safety of missions. The task goals for such evaluations affect evaluation weighting, and these scores change criteria like the duration of the mission and the promptness. Risk matrices may also be used for assessing the status and effect of possible threats within a system. Traditional risk evaluation matrices classify risk with magnitude and frequency and futile reasoning to expand the potential beyond such classifications.

Crown Jewel Analysis contributes to evaluating and informing more risk analyses in kinetics, cyber, and supply chain, leveraging cyber assets that are mission-critical for the process. The project goals for these evaluations affect the relative weight of the review, with considerations like mission duration and directness change. Implementations for risk assessments help companies identify their potential weaknesses and determine their potential impacts on the enterprise and its processes. The CSCRM is described as the organizational approach and programmatic practices used to evaluate and manage risks over the whole IT supply chain, hardware and software supply chain processes. Testing strategies involve deciding whether the network security objective of the enterprise is to avoid an attack, respond appropriately when one occurs, or both.

The supply chain development responses of Pettit, Fiksel, and Croxton have three possible end states which decide that if a risk is unreasonable, the efficiency is enhanced, or productivity degraded, based on the system's potential risk situation. Capacity considerations that may not always be numerical in military supply chains should be included in a risk evaluation of factors such as reconstruction, capacity, versatility in procurement, and total capacity that impact organizations.

Due to increasing cyber risk sophistication frameworks over the past decade, the threats to the future of new cyber security incident management and harm risk mitigation approaches and concepts are apparent. The transitions in technologies, processes, and network configurations also need innovative strategies to resolve and minimize the influence of cyber safety incidents. Unique, agile, and integrated market systems like supply chains are not a marginal method or an active analysis area in current and evolving frameworks.

6.9.2 Models of Risk Management

Existing supply chain management models should be applied in enterprises to enhance supply chain operations. First, it defines process management goals and then measures, analyzes, improves, and ultimately controls the process. The Six Sigma DMAIC model included in 2006 decided that although the technique could also be used, there were limiting factors which limited their likeliness to be implemented, including stock keeping policies and the degree of activity, in an examination of the adequacy of implementing the Six Sigma methodology to the UK supply chains. The Cyber Kill Chain is a way to define the phases of an assault in cyberspace. The Cyber Kill Chain phases include recognition, armament, supply, use, activation, command and control, and targeting operation. Re-evaluations of the cyber physical system model have suggested modifications to the well-established interpretation of attacks, particularly within communication systems and computer objects with concrete results. The SCOR is a cross-industry platform to evaluate and improve the efficiency and management of the business across the whole company. It is an option in the Cyber Kill Chain. SCOR illustrates and interconnects the following specified modules, procedures, benchmarking methods, management strategies, and application navigation. SCOR aims "to collaborate, compare, and establish new or enhanced processes in the supply chain" [9].

6.9.3 Existing Technology of Risk Management Models

The possible risks generated by deploying advanced technologies in military supply lines may be evaluated using risk identification and organizational design. Many frameworks are available, namely CSCRM, management of credit risk, Crown Jewel Analysis, and the supply chain's durability architecture. While these models contribute to the risk evaluation, they do not offer a systematic and balanced assessment of the direct and indirect factors required to use emerging technology in military supply chains. Many attempts were made to simulate processes and frameworks in the semitone risk evaluation, with some success. Considering the scope, nuance, and meaning necessary for the various businesses, governments, and political areas, this is an ongoing field of research and development, which adds more sophistication, visualization, and simulation required by cyber application into these systems.

Existing network security management and supply chain management models are primarily based on cyber security or Supply Chain 4.0. The models under discussion were cross-compatible, but none directly established the strategic background of supply chains. Although components of these frameworks will help explain how emerging technologies can affect operational supply chains, they are not an overall solution, and the study gap persists.

6.10 Latest Technological Models Used for Supply Chain 4.0

Multiple attack vectors make the protection of supply chains insecure, and therefore, these vectors should all be regarded to create resilience to the defensive supply chain. The field of technology is constantly expanding. This is shown by Moore's rule, which makes computer boards more economical and accessible. Since a comprehensive model for evaluating the effect of any technology on military supply chains cannot be found, this section involves reviewing purely technical models to determine the main characteristics that could apply to a technical process framework. The model emphasis on Supply Chain 4.0, the military background, and the aim as an evaluation system would be the main criteria for this evaluation.

6.10.1 Latest and Effective Wireless Communication Systems

The fourth generation of wireless connections is crucial to boost internet, Industry 4.0, and related global supply chains (5G). 5G aims to provide heterogeneous networks with high speed, cheap networking by using more excellent range bands. By introducing 5G, it would be possible to access the necessary technology for broad track and trace networks, such as those used for defense supply chains. A better abundance of data and the introduction of potential risks for IoT applications in supply chains falls along with the delivery of 5G networks. The invention of "monitors and locates," the ability to track items in transit and inventory in real time, is one of the critical cases for 5G. Tracking gives competitors less ability to poison the supply chain or cut supply chain output to benefit their competitiveness. However, introducing such technology could also have possible consequences requiring further testing. Track and trace enable products to be monitored in supply chains in virtual environments. This may have significant tactical effects in a military setting, with such devices likely to be dependent upon in operation operations, such as the supply of fuel and munitions. Monitor and tracking enable precise forecasts and product development and enhance transparency inside the supply chain. Any application integrates 5G with RFID tags or related systems to label artifacts that move through networks.

6.10.2 Threats in Wireless Communication Systems

The 5G technology depends on significant, current, capacity-based functionality. Defense and combat technologies can be implemented in areas that exist or are disputed. Operating without independently run networks presents additional obstacles, as does the idea of owning such a system that possible opponents regulate.

RFID infrastructure-focused cyber invasions often attack anonymity, encryption, or access to trace and detect systems and infrastructure. Via acts like sensitive monitoring and inventory records, these assaults can be embodied by creating illegal identity tags and preventing access by lawful vendors to RFID facilities, where appropriate [12]. The remoteness, the abuse of the products and the facilities of the chains, and their redirected commodities/services are all possible threats; they include strategic nuclear bases.

Fortunately, the 5G technology has shortcomings that influence its defense track and trace features. 5G suffers from a substantial lack of insertion from multiple input and numerous electrical technologies to work at necessary concentrations. These standards face possible problems in operating military settings where there might not be an acceptable degree of facilities. Lower generations in telecommunications, reduced data transmission rates, and perhaps harmful evolution of an organization may also have to be relied upon.

6.10.3 Emerging System in Wireless Communication Systems

The bridge of new systems from many other sectors of computer hackers, such as cryptographic, has generated contemporary solutions to handle these issue areas. However, industrial RFID tags can lack the needed computational capacity to support public-key cryptography due to their high complexity, poor efficiency, and high cost. As the 5G technologies enable the material to be simultaneously tracked and traced, challenges to these networks can impair their military capability in real time. This is interdisciplinary. Many research works are minor in scale, and their conclusions do not apply to other sophisticated fighter supply chain technologies, particularly in light of the consequences of 5G adoption.

To impose on all suppliers the ubiquitous deployment of the RFID tracking and tracing into the supply chain, the US Defense has appealed the decision of activated RFID tag requirements. This practical application presented issues throughout the integration phase, showing an insufficient grasp of how technologies may affect the supply chain and the need for more implementation evaluations.

6.11 Cloud Computing CPS Supply Chain Management

Cloud computing applies to "the technologies provided in the data centres providing such resources as services on the internet and to hardware and systems software." Millions of people worldwide, infrastructure, automation, network access, and storing elements are used as components of cloud computing, found in application groups such as service applications, a database, and networks as a service. Business opportunities for technical solutions such as fog-2-fog collaboration exist in developments in such areas as fog and cloud technology that offer cloud-specified prominent features. Cloud computing generally benefits from simplicity, place isolation, scalability, and cost-efficiency. But cloud infrastructure also creates new challenges for IoT devices in terms of stability. In the new arena, most of the audience search for different techniques in the market. They are finding search exciting concepts, one of which is CPSSCM. According to Table 6.1, in year 2020, most people used the highest web search about CPSSCM but in 2008 more minor arch than in the year 2010 and again increase in 2014 but year 2020 shows significantly less search on

TABLE 6.1

Search about CPS Supply Chain Management

Year	Image search: all category	Web search: all category	YouTube search: all category
2008	494	633	385
2009	497	572	334
2010	471	535	464
2011	462	525	523
2012	466	493	485
2013	451	475	631
2014	472	475	623
2015	477	471	807
2016	513	475	891
2017	646	485	781
2018	839	503	521
2019	866	525	561
2020	965	539	494

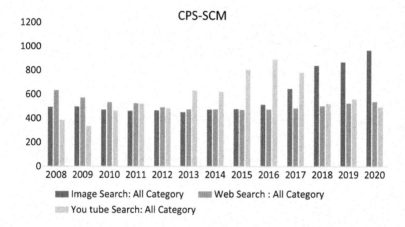

FIGURE 6.3
CPSSCM.

CPSSCM. CPCM-related concept was YouTube's highest search in 2016 and was substantially less searched in 2009. Image search on CPSSCM was highest in 2020 and significantly less in 2013 as shown in Table 6.1.

Data set: Data is taken from Google Trends from 2008 to 2020 for 11 years for different categories, as given in Table 6.1.

Figure 6.3 shows year-wise categories searched by users. The single type shows the highest density compared to others.

6.11.1 Threats against Cloud Computing CPS Supply Chain Management

Cloud networking weaknesses include service reliability problems, memory allocation mistakes, location-based problems, application control issues, overflows of authentication queries, and service denial of assaults. Data routing between IoT sources and the cloud service providers that handle the data is based on secured, private interfaces. Identify innovative for a range of IoT wearable sensors, such Wirelesshart, ZigBee, IEEE 1451, and 6LOWPAN might differ across those protocols. One suggested approach for improving the compatibility of various IoT sensors and devices is the intelligent gateways and the fog computing architectural design; this is achieved by investing additional capital into IoT color's architecture.

Because cloud services combine large data stocks from various networks in a highly virtualized environment, identity management is a significant obligation for providers. Clouds are essentially resource sharing by multitenancy schemes, ensuring that multiple users are not isolated at the hardware level. Multitenancy can create barriers to secrecy, as records can be broken and divulged, and remains as allocations for information sharing shift. One way to reduce computer characteristics is to protect messages and then delete the passwords so they can't be retrieved even though it's accessed later.

Another cloud problem for IoT is its availability. IoT innovations using the cloud have to have daily access to those resources for the provision and acquisition of data. Then the performance of affected IoT devices could be reduced as clouds for IoT go offline or become inaccessible. The design of cloud technology allows the generation of reliable failing patterns challenging for unit commitment. That may mean unpredictable service drop-outs, and several hours of accessibility can be prevented. Thanks to the omnipresent incorporation of modern IoT technologies into certain parts of society, IoT cloud interruption can have a devastating impact on consumers, for example, by preventing them from communicating with protection or security systems.

Denial of service assaults pose a unique challenge to IoT-enabled cloud computing platforms, as modern devices are increasingly embraced. While cloud servers are versatile enough to change resource specifications, they cannot deny service attacks. Decentering the cloud is one method of reducing network denials, but it may come at the cost of other safety hazards like reduced resource elastic modulus and more specific attacks.

6.11.2 Strategies in CPS Supply Chain Management in Cloud Computing

Cloud infrastructure aims to incorporate into strategic supply chains as contractors use these resources more and more to reduce cost and performance. If cloud computing begins to impact companies, government institutions, and people interconnected technological environments, it is crucial to understand how cloud computing services can impact those participants.

Fog computing is another basic technique for Industry 4.0. Fog computing is a grid computing model that applies to cloud computing closer but nearer to the edge. The fog will handle time-sensitive activities and other connected elements to the server, securing the cloud benefits closer to the data sources. The low latency results for Industry 4.0 have some tangible advantages. Fog computing was considered and evaluated in situations leading to UAV integration into catastrophe scenarios. This study examined the effect of fog expansions in disaster scenarios such as flooding at the edge of the cloud and how these technological developments can improve service quality (QoS) in predicting and preparing catastrophes. In assessing the appropriateness of these fog deployments, the research highlights many considerations, including technological needs such as intensity and range, cooperation with nodes and teamwork criteria, including latency and confidence. While the report does not concentrate directly on military supply chain situations, this study aims to determine how emerging technologies such as fog can affect current systems.

The fog computing trust management (COMITMENT) improves service quality and historical metrics to boost fog architectural security as a cloud extension. However, the approaches outlined in the methodology are intended to have quantitative and diverse implementations in different contexts. This research does not rely on the military or supply chain.

Tariq proposes a Hybrid Cloud Computers' Agent-based Information Security Program that serves as a decision-making mechanism to evaluate threats within a cloud infrastructure and respond to them. The job is highly risk-oriented and has no account of implications in this sense that goes beyond properties, risks, and vulnerabilities. The research should not even concentrate on possibilities for the defense chain. The study results, however, help to determine the efficacy of various risk management approaches in the context of taking into account risks to cloud service systems.

6.12 IoT and Industry 4.0 Systems in CPS Supply Chain Management

The IoT refers to the internet-based computer connections that provide physical world information through observation, like sensors. IoT allows data from reality to be collected and promotes customer perception and a willingness to adapt to situations. IoT supplies computing and internet traffic to current computers that can connect heterogeneously through networking devices. In the production and sale of products, the IoT is also used. IoT systems have been credited with up to 20 per cent improvement in industrial performance. More and more networks are used to operate services within industry-wide communications systems, including supporting technology such as wireless sensing and cloud usage.

The IoT is an enabling infrastructure for the new Industry 4.0 idea where innovative production processes are developed. Other technologies, such as cloud, extensive data analysis, and cyber-physical networks, are also supported by Industry4.0 in addition to IoT. Industry 4.0 offers fantastic things, on-demand production, energy-efficient homes, intelligent cities, and technology in all aspects of everyday life. The IoT allows data to be collected from reality that promotes a change in the view of consumers and a desire to personalize the situation. The technology which allows IoT extension include RFID tagging (RFID), sensors, Near Field (NFC), cloud computing, Wi-Fi and Cellular networking. The modern features are also available. IoT is now integrated into the production and delivery of goods, allowing for the creation of personalized and scalable products.

The growth and continuous enhancement of demand production is a complementary technology for Industry 4.0. Highly tailored, scalable production and supply chain management systems rapidly permit re-tooling and system customization throughout the near future. The linkage with 3D printing, cutting tools, and other processes managed by computing enables improvements in their production and supply processes. 3D printers can generate mobility in supply chains that allow the machinery to be printed and replaced locally. The 3D printing of instruments on Earth's Space Station has illustrated this principle. The enhanced network access that would enable remote publishing and object formation enhance the scope of a cyberattack. The skills and threats related to supply chain management would change with 3D printing technology. With seamless integration, agile supply chains are being interrupted.

In addition to converting military customers into "prosumers" as manufacturing consumers, the introduction of Industry 4.0 in military supply chains can allow agile, fast response supply, and product growth. Industry 4.0 offers defense, as the equipment can be modified more quickly and efficiently to its environmental requirements, and logistical assistance can be prepared more reliably. Industry 4.0 will transform how goods are produced within the framework of the military supply chains. For Industry 4.0, future dynamic network designs will include the durability and flexibility required to operate the supply chain in agile military ecosystems. In addition, however, Industry 4.0 has some possible security problems, and the systematic consequences of cyber protective action violations are not well understood for Industry 4.0 deployments.

6.12.1 CPS Supply Chain Management Threats to IoT and Industry 4.0 Systems

The present challenges of IoT research include uniformity, safety problems, and the possibility of data leakage. Obstacles to omnipresent internet connections, such as connections security limits, operational considerations, and network abandonment, are susceptible to seriously disrupting the conduct of armed forces, particularly in the field. Highly classified defense information is also a threat to IoT deployment, as it is likely that computers will not capture security information without authorization. IoT adoption has many variables that minimize deployment viability in the military background and IoT adoption. In armed force environments, IoT technology must be strategically applied and leveraged.

IoT devices depend on a network infrastructure outside of themselves, like analytical platforms. This puts constraints on the isolation of military capabilities and resources from commercial networks and thus enhances the military's trust in third-party technology service providers. Furthermore, these third parties and their control of existing infrastructures will directly affect the service life of those systems used by armed services. This raises further concerns regarding the competence of these devices as military assets and the changes and contacts that third-party facilitator can make to them.

The integration of IoT and Industry 4.0 ecosystems of fog computing has both safety consequences and can also improve the safety of military environments. There is no guarantee of stable cloud connection; fog technologies may be deployed in locations where there is little to no signal. However, some architectural advantages emerge, bringing additional efficiency and safeguards. There are also solid semantical areas of research aimed at defining and contextualizing the field in this area and related areas.

The processing, owing to operational needs for in-house data and technical specifications for the external manufacturing process, filtering, and analytics (DPS) of the IoT sensors and devices, often presents problems for the military. It is not practicable to gather data and submit it to be stored abroad in some situations by providing it before restoring the data to the original site. This needs reliance on multiple networks, both secure and public, encryption across a dynamic sequence of events; confidence in the protection of third parties; data storage and analysis for data mining. In designing products according to the requirements, the interconnectedness of small production plants and the capacity to reprogram and modify production and industrial products in real time also play a key role. Interconnection and versatility are two essential characteristics of product growth that affect product design in Industry 4.0. Production advantages on request and Industry 4.0 innovations have to be measured at a granular level.

6.12.2 Decision-Making in IoT and Industry 4.0 CPS Supply Chain Management

To measure the impact of the IoT on the distribution network, we combined the "neutrosophic decision-making formulation and confirmation technique (N-DEMATEL) approach with analytical hierarchy (AHP)". AHP determines the cause and development of the safety conditions of a supply chain using graded parameters. The DEMATEL technology was used to cover many situations in the supply chain, including the replacement parts industry, environmentally sustainable supply chain management, and the rapid turn-around of consumer items. This method also can be used in military supply chain situations with different technologies but is probably to take account of the particular features of supply chain defense chains such as in tactical settings in a broader model. Cyber risk evaluations for IoT products were often suggested as methods to measure the effect of IoT technology on military supply chains.

6.13 Conclusion

A morphological box is presented in this chapter to design the CPS in SC planning and control processes. A literature review to classify 11 traits of distinct expressions is carried out for this reason. These are classified into the features of method and technology. In addition, a steel manufacturing firm applied morphology in two usage cases. In brief, the technological characteristics deal with the advancement in the structural architecture of a CPS. In SC preparation and control systems, adding method features into the morphological box also considers the criteria for these system solutions. The process owners also take part in an early CPS design stage.

References

1. Sharma, A.; Jain, D.K. *A Roadmap to Industry 4.0: Smart Production, Sharp Business and Sustainable Development*; Springer, Cham, Switzerland, 23–38. 2020.
2. Mentzer, J.T.; DeWitt, W.; Keebler, J.S.; Min, S.; Nix, N.W.; Smith, C.D.; Zacharia, Z.G. Defining supply chain management. *J. Bus. Logist.* 22(2):1–25. 2001.
3. Frederico, G.F.; Garza-Reyes, J.A.; Anosike, A.; Kumar, V. Supply chain 4.0: Concepts, maturity and research agenda. *Supply Chain Manag. Int. J.* 25(2):262–282. 2019.
4. Waters, D.; Rinsler, S. *Global Logistics: New Directions in Supply Chain Management*; Kogan Page Publishers, London, UK, 100–127. 2014.
5. Ivanov, D.; Dolgui, A.; Sokolov, B. The impact of digital technology and industry 4.0 on the ripple effect and supply chain risk analytics. *Int. J. Prod. Res.* 57(3):829–846. 2019.
6. Garrido-Hidalgo, C.; Olivares, T.; Ramirez, F.J.; Roda-Sanchez, L. An end-to-end internet of things solution for reverse supply chain management in industry 4.0. *Comput. Ind.* 112:103–127. 2019.
7. Moustafa, N.; Adi, E.; Turnbull, B.; Hu, J. A new threat intelligence scheme for safeguarding industry 4.0 systems. *IEEE Access* 6:32910–32924. 2018.
8. Salamai, A.; Hussain, O.K.; Saberi, M.; Chang, E.; Hussain, F.K. Highlighting the importance of considering the impacts of external and internal risk factors on operational parameters to improve supply chain risk management. *IEEE Access* 7:49297–49315. 2019.
9. Dhatterwal, J.S.; Kaswan, K.S.; Preety. "Intelligent agent-based case base reasoning systems build knowledge representation in Covid-19 analysis of recovery infectious patients" in a book entitled "application of AI in COVID 19" published in Springer series: Medical virology: From pathogenesis to disease control, July 2020, ISBN No. 978-981-15-7317-0 (e-Book), 978-981-15-7316-3 (Hard Book). https://doi.org/10.1007/978-981-15-7317-0.

10. Turnbull, B. Cyber-resilient supply chains: Mission assurance in the future operating environment. *Aust. Army J.* 14(2):41–56. 2018.

11. Keshk, M.; Sitnikova, E.; Moustafa, N.; Hu, J.; Khalil, I. An integrated framework for privacy-preserving based anomaly detection for cyber-physical systems. *IEEE Trans. Sustain. Comput.* 6: 66–79. 2019.

12. Marsden, T.; Moustafa, N.; Sitnikova, E.; Creech, G. Probability risk identification-based intrusion detection system for SCADA systems. In *International Conference on Mobile Networks and Management*; Springer, Berlin/Heidelberg, Germany, 353–363. 2017.

13. Martin, C.; Towill, D.R. Supply chain migration from lean and functional to agile and customised. *Supply Chain Manag.* 5(4):206–213. 2000.

14. Boyson, S. Cyber supply chain risk management: Revolutionizing the strategic control of critical IT systems. *Technovation* 34(7):342–353. 2014.

15. Xiao, Q.; Boulet, C.; Gibbons, T. RFID security issues in military supply chains. In S. Vyas, V. K. Shukla (eds.) *Proceedings of the Second International Conference on Availability, Reliability and Security (ARES'07)*, Vienna, Austria, 599–605. 2007.

16. Hendricks, K.B.; Singhal, V.R. Association between supply chain glitches and operating performance. *Manag. Sci.* 51(5):695–711. 2005.

17. Liya, J.; Tiening, W.; Ronghui, W. Risk evaluation of military supply chains based on the case and fuzzy reasoning. In T. Sobb (ed.) *Proceedings of the 2010 International Conference on Logistics Systems and Intelligent Management (ICLSIM)*, Harbin, China, 1, 102–104. 2010.

18. Mei, M.M.; Andry, J.F. The alignment of business process in event organizer and enterprise architecture using TOGAF. *JUTI J. Ilm. Teknol. Inf.* 17(1):21–29. 2019.

19. Zarour, K.; Benmerzoug, D.; Guermouche, N.; Drira, K.A. A BPMN extension for business process outsourcing to the cloud. In *World Conference on Information Systems and Technologies*; Springer, Berlin/Heidelberg, Germany, 833–843. 2019.

20. Srinivasan, S.; Singh, J.; Vivek, K. Multi-agent-based decision support system using data mining and case-based reasoning. *Int. J. Comput. Sci. Issues* 8(4) No 2, July 2011.50 ISSN (Online): 1694-0814.

21. Scrapper, C.; Droge, G.N.; Xydes, A.L.; de la Croix, J.P.; Rahmani, A.; Vander Hook, J.; Lim, G. *Mission Modeling Planning, and Execution Module (M2PEM) Systems and Methods*. US Patent Appl. 16/403, 838. 2019.

22. Sampath Kumar, V.; Khamis, A.; Fiorini, S.; Carbonera, J.; Olivares Alarcos, A.; Habib, M.; Olszewska, J.; Li, H.; Olszewska, J.I. Ontologies for industry 4.0. *The Knowl. Eng. Rev.* 34:E17. 2019. https://doi.org/10.1017/S0269888919000109.

23. Waters, G.; Blackburn, A.J. *Australian Defence Logistics: The Need to Enable and Equip Logistics Transformation*; Kokoda Foundation Limited Publisher, Number 19: Balmain, Australia. 2014.

24. Wang, F.; Ge, B.; Zhang, L.; Chen, Y.; Xin, Y.; Li, X. A system framework of security management in enterprise systems. *Syst. Res. Behav. Sci.* 30(3):287–299. 2013.

25. Flauzac, O.; González, C.; Hachani, A.; Nolot, F. SDN based architecture for IoT and improvement of the security. In M. A. Abid (ed.) *Proceedings of the 2015 IEEE 29th International Conference on Advanced Information Networking and Applications Workshops*, WAINA '15, Gwangju, Korea, 688–693. 2015.

26. Poudel, S. Internet of things: Underlying technologies, interoperability, and threats to privacy and security. *Berkeley Tech. L.J.* 31:997–1022. 2016.

27. Seliem, M.; Elgazzar, K.; Khalil, K. Towards privacy-preserving IoT environments: A survey. *Wirel. Commun. Mob. Comput.* 1032761. 2018.

28. Shon, T.; Cho, J.; Han, K.; Choi, H. Toward advanced mobile cloud computing for the internet of things: Current issues and future direction. *Mob. Netw.* 19(3):404–413. 2014.

29. Pathan, A.S.K.; Lee, H.W.; Hong, C.S. Security in wireless sensor networks: Issues and challenges. In *Proceedings of the 2006 8th International Conference Advanced Communication Technology*; Phoenix, Park, Korea, 44(2):1043–1048. 2006.

30. Aazam, M.; Huh, E.N. Fog computing and smart gateway-based communication for the cloud of things. In MDPI and R. Mohmmad (ed.) *Proceedings of the 2014 International Conference on Future Internet of Things and Cloud*, Barcelona, Spain, 464–470. 2014.

31. Singh, J.; Pasquier, T.; Bacon, J.; Ko, H.; Eyers, D. Twenty security considerations for cloud-supported Internet of things. *IEEE Internet Things J.* 3(3):269–284. 2016.
32. Mendes, R.; Vilela, J.P.; Privacy-preserving data mining: Methods, metrics, and applications. *IEEE Access* 5:10562–10582. 2017.
33. Saleh, M.; Khatib, I. Throughput analysis of WEP security in ad hoc sensor networks. In STM Journal and B Balaji Bhanu's *Proceedings of the Second International Conference on Innovations in Information Technology (IIT'05)*, Dubai, UAE, 26–28. 2005.
34. Awasthi, A.; Grzybowska, K. Barriers of the supply chain integration process. In A. Awasthi *Logistics Operations, Supply Chain Management and Sustainability*; Springer, Cham, Switzerland, 15–30. 2014.

7

Security and Privacy Aspects in Cyber Physical Systems

Himanshu, Bharti Nagpal, Ram Shringar Rao, and Shobha Bhatt

CONTENTS

7.1 Introduction

Cyber physical systems refers to modern systems with a combination of physical and computational abilities which can interact with humans in a new way. Expanding and interacting with the physical world by methods such as communication, control key, and computing are the primary foci of current and future study [1]. And these research challenges include design and implementation challenges. Engineering norms centered on technology and supported by solid mathematical assumptions form the basis of cyber physical systems. Both the physical and digital components of pervasive computing may benefit from the notion of privacy and security. Information assurance, information protection, and cyber security are

DOI: 10.1201/9781003220664-7

just a few examples of phrases that mean essentially the same thing but sound different. The informational and non-physical aspects are crucial to the function of cyber physical systems. Cyber physical system is a vast tree that includes Internet of Things, robotics, machine automation, process control system, industrial internet, and industrial control systems. The main five principles for the nonphysical side of cyber physical systems are confidentiality, integrity, availability, authorization, and nonrepudiation. There are so many methods available for implementing these principles. In order to ensure that the data has not been tampered with, for instance, an encrypted data and block cipher are utilized. Confidentiality, in this context, is keeping sensitive information hidden via the use of cryptography and decryption methods. The redundancy function is used to ensure that all necessary system components are always up and running. The authentication principle ensures that only authorized users may get access to a resource by using tools like passwords and certificates. The diabetes and diabetic principle follows from the authorization concept and states that once a message has been sent or an approved use of an item has taken place, neither may be revoked. Privacy is the word that is related to the confidentiality principle. The interconnection of elements in the cyber physical systems leads to regular inspection. Because the cyber physical systems is the combination of both cyber and physical processes which means it gives the pathway to both the physical and cyberattack. The vulnerabilities that may occur on the physical side and cyber side can't be added they are multiplied. To achieve a fully protected cyber physical system the protection is to be done both on the cyber side and the physical side. Because protecting a single pathway doesn't protect the whole cyber physical system. Furthermore, the potential components may be safeguarded by the usage of elements that are employed in both the physical and cyber realms. In a cyber physical system assault, the attacker goes at the devices themselves, the interconnection between them, the CPS's supporting facilities, and the internet. The attacker uses the ambiguity in the weak inter-protocol communications to their disadvantage. They also make use of security flaws in the poor design and the implementation of Application Program Interface (API) to trade off the components. They also monitor the communication between the clients and the peer-to-peer communication between devices. So each of the vulnerabilities is covered by the security and also secures the components [2].

Aeronautics is one industry that makes extensive use of cyber physical systems for things like aircraft landings, flight crew communications, and flight health monitoring. Firefighters, farmers, and those who clean up after accidents like mudslides or chemical spills at sea may all benefit from cyber physical systems. Cyber physical systems are also used in the medical and transportation sectors.

Section 7.2 contains the literature review of eight papers. Section 7.3 presents the general architecture of cyber physical systems. Section 7.4 presents the proposed framework of cyber physical systems (CPS) in the transportation system. Section 7.5 contains the security and privacy issues in cyber physical systems along with their countermeasures. Attacks on the cyber physical systems are discussed in Section 7.6. Section 7.7 contains the advantages and disadvantages of cyber physical systems. An application of cyber physical systems is discussed in Section 7.8 and then Section 7.9 contains the conclusion of this chapter.

7.2 Literature Review

Jianuha shi, jiafu wan, and Hehua yan hui suno (2011) elaborate on the survey of cyber physical systems. The cyber physical systems and its features are also elaborated on by the authors. The authors illustrate the layout of cyber physical systems and also discussed the research process. The research processes include energy control, security control, management and transmission, and model-based software design. The classic applications of the cyber physical systems are also illustrated in this paper. The present condition is contrasted to the traditional trend of features, and the characteristic research issue is examined [3].

Jairo Giraldo and Esha Sarkar et al. (2017) describe security and privacy issues in cyber physical systems in a survey of studies. This paper elaborates the general cyber physical systems diagram and the taxonomy related to CPS. The CPS security and privacy taxonomy which are domain, security, privacy, and defenses are also illustrated by the authors. The authors also described the taxonomy-related work with the help of graphs to understand which taxonomy is illustrated in which paper. The authors also gave the future research direction and the recommendation in the future research area. The taxonomy-related papers are selected and presented [4].

Abdul Malik Humayed et al. (2017) describe cyber physical systems security in a study. The three coordinates of the cyber physical system security with the help of a framework are described in this paper. These three coordinates are cyber physical systems, threats, and possible vulnerabilities. The authors also presented the applications of the cyber physical systems with the help of diagram authors described all the applications. The threats against the cyber physical systems that are illustrated in this paper are grid-related threats, ICS-related threats, medical devices-related threats, and smart cars-related threats. The vulnerabilities that lead to these threats illustrated above are also defined in this paper. The authors also described the attacks on the CPS applications and also the security controls related to them. The challenges related to the cyber physical systems are also described in this paper [5].

K.P. Chow, Eric Ke Wang, Xiaofei Xu, S.M. Yiu, Yunming Ye, and L.C.K. Hui (2010) discuss the security issues and challenges for cyber physical systems. This paper presents the basic workflow of cyber physical systems. The general workflow includes four steps which are monitoring, actuation, networking, physical process, and computing. The confidentiality, availability, integrity, and authenticity of security objectives are also described in this paper. Important types of attacks are also defined in this paper and these attacks are denial of service attack, key compromised attack, eavesdropping, and man-in-the-middle attack. The authors also presented the importance of the context-related framework and its working in this paper. The workflow of that framework is also illustrated in this paper with the help of adversaries of the security [6].

Francesco Bullo, Fabio Pasqualetti, and Florian Dörfler (2013) elaborate on the attack identification and detection in cyber physical systems. This paper presents a mathematical structure for attack identification, detection, and monitoring. The authors also discussed the fundamental parameters and limitations regarding the monitoring process. The graphical representation of classic and theoretical characteristics is also presented in this paper. The authors also design the monitors for attack detection and identification. With the help of the example, the result is validated and verified. The basic monitoring limitations include system-based theoretical monitoring and graph-based monitoring limitations. Certain examples for understanding the concept are also presented in this paper [7].

Ragunathan (Raj) Rajkumar, insup lee et al. (2010) describe the cyber physical systems as the next computing revolution. This paper elaborates on how the cyber physical systems are used in companies nowadays. The grand challenges and the vision towards the future are also discussed in this paper. The authors also illustrated the foundation and challenges related to the scientific approach in cyber physical systems. The various problems in cyber physical systems are safety, robustness, composition, architecture, hybrid systems, computational transparency, and sensors. The authors also discussed the infrastructure and the social impact of cyber physical systems [8].

Zhipeng Cai and Xu Zheng (2018) elaborate on the mechanism for uploading data in the cyber physical systems. Firstly, the authors described the problem statement which consists of functions like system input, utility, privacy, and the design objectives.

This article also presents a framework for uploading data, detailing the many processes necessary to carry out this operation, including determining the operation's difficulty, providing a comprehensive overview of the framework, and choosing a content-based algorithm. This research also presents an examination of productivity while using a greedy method with a trace opposite shoulder. All the performance analysis factors like time complexity, effectiveness, and completeness are also calculated in this paper based on the performance. The evaluation of the performance is also presented in this paper by taking general performance and performance with the individual. With the help of the graph, all the performance is shown in this paper [9].

Jinjun Chen, Muneeb Ul Hassan, and Mubashir Husain Rehmani (2019) elaborate on the distinctive privacy issues in cyber physical systems. This paper presents comprehensive privacy issues in cyber physical systems applications like transportation system, energy systems, power grid, and healthcare systems. The comparison table based on the privacy strategies like encryption, anonymization, and differential privacy is also presented in this paper. A framework related to the differential privacy in cyber physical systems is also presented in this paper. The authors also compare the design parameters of the cyber physical systems with the help of a table so that it can be easily understood. The authors basically taking the application one by one and then with the help of framework and diagram they elaborates the whole process and then with the help of table and taking certain parameters for the comparison for a single application of cyber physical systems [10].

7.3 Architecture of Cyber Physical Systems

Figure 7.1 depicts the overarching structure of cyber physical systems as a four-tiered architectural; at the lowest level, embedded with sensors measure and change physical quantities, respectively. After the output of the sensors and actuators has been measured and modified, the human operator is utilized to do any remaining network- and sensors-related tasks, such as assessment. The application layer is between the network layer and the physical layer. The third layer is the network layer which connects the physical process to the internet which is also known as the cyber part of the cyber physical systems.

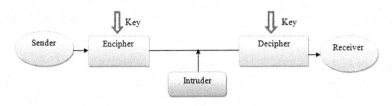

FIGURE 7.1
Basic architecture of cyber physical system.

The network layer generally provides the interconnection between controllers to the user. The network layer is a part of the cyber part of the cyber physical systems. So it is a very important layer in this architecture and most of the operation is done by that layer. The fourth layer is a hardware and software control layer in which all the controlling related information related to sensors and actuators is taken by this fourth layer hardware and software control layer. It is the brain of the cyber physical systems because all the important problems are processed by that layer.

7.4 Proposed Framework of Cyber Physical Systems in Transportation System

Figure 7.2 describes a suggested architecture for a cyber physical system in the transportation sector, where data from three types of vehicles (electric cars, autonomous vehicles, and trains) is stored in the cloud and queried by an analyst. Increases in efficiency and convenience for both motorists and passengers are driving the transportation industry's steady expansion. Railways are the main mode of freight transport around the world, so the advancement in railways has brought several opportunities and challenges. As the advancement is also done in the vehicular networks so it makes possible for vehicle to vehicle connections and vehicle to drivers connections. But the connections and the cloud storage is little prone to the attacks, but it have more secure than the previous framework

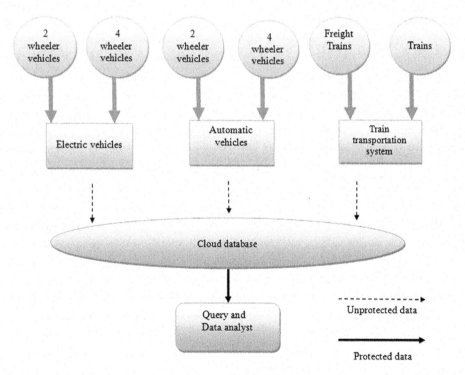

FIGURE 7.2
Cyber physical system in transportation system.

for the transportation system because here the protected data and unprotected data is classified so that the main focus is to be done on the data which is more important. Some privacy and security also added at the end of the cloud database so that the storage data is found more protected by the analyst. Encryption and hashing is used during the communication of vehicles and drivers so that the intruder does not take the advantage of that particular communication channel.

7.5 Security and Privacy Issues in Cyber Physical Systems along with Its Countermeasures

Control security and data security are the two main areas of CPS security. Data security refers to the practice of protecting data throughout its transmission, storage, and use in a network, especially in the case of ad hoc or other types of networks with a high degree of loose coupling. The control security involves rectifying any command-related matters in the network environment and lightens the command structure from any type of attack on the control methods and system evaluation [11]. Data security's primary function is to prevent unauthorized access to data by use of cryptographic safeguards. The most important factors affecting information integrity are as follows:

1. Secured access to device
2. Secured transmission of data
3. Securing the applications
4. To achieve secure data storage
5. Secured actuation and sensors

7.5.1 Risk Assessment

Risk assessment is an essential element in the security aspects of CPS. With widespread use, cyber physical systems (CPS) have become an easy target for hackers. There has to be a shift in security's primary focus, hazard identification, from the evaluation of individual computers to that of whole networks. The primary goal of this strategy is to ensure the safety of future cyber physical systems. The cyber physical system risk assessment technique involves three steps, which are specifying what might occur to the systems, estimating the chances of the incident, and assessing the results. There are three more components while doing CPS risk assessment which are listed below:

1. Asset recognition
2. Threat recognition
3. Vulnerability recognition

7.5.2 CPS Security Analysis

When a gravitational acceleration is combined with a digital one, like in a cyber physical system, new difficulties arise and must be taken into account throughout the system's design. Adding to the difficulty is the fact that further protections must be taken to protect private information from the myriad of devices that may connect to the network at any one time, regardless of their location. For avoiding challenges during designing, the designer should include these three parameters in the mechanism that is detection, mitigation, and prevention. There are some attackers who do not attack the direct vulnerability but try to attack the layers vulnerability simultaneously. So some phase which might kept in mind during the designing are listed below:

1. Security analysis on perception layer
2. Security analysis on transmission layer
3. Security analysis on application layer [12]
4. Network access security
5. RFID and WSN security analysis

7.5.3 CPS Security Solutions

A different type of security is required at different stages of cyber physical systems. For example data privacy is an important concern of the intelligent transportation system and medical intelligent medical system. Data authenticity is more important in the smart grid and intelligent urban management. There are so many researchers who have proposed the framework and models for the cyber physical system. Some security techniques are also presented and addressed in the following subsections. The two subsections are the single-layer-based solutions and multilayer-based solutions.

7.5.3.1 Single-Layer-Based Solutions

In order to improve encryption methods, a researcher suggested an identity-based distribution channels, while another author provides key management encryption [13]. So it is illustrated that if the small key size is used then the computational process also takes less time in the asymmetric key encryption [14]. If the available sources are limited then it will also do the cost estimation analysis. For preventing the RFID tags from the sniffing attack a lightweight authentication protocol is proposed. This protocol also provides mutual authentication for the RFID readers [15]. The authors also proposed the secured WSN that combines the healthcare system and cloud computing. The proposed framework also does the monitoring and decision-making [16]. For providing message integrity, authorization, and authentication, a two-step mutual scheme is proposed for the smart grids. Diffie-Hellman algorithm is used for achieving the session key exchange [17].

7.5.3.2 Multilayer-Based Solutions

Some writers have provided multilayer-based solutions for cyber hardware and software confidentiality and protection, recognizing that single-layer approaches to cyber security problems are not always enough. For example, securing just one layer of a cyber physical system won't make it bulletproof; instead, you will need to protect all of the layers. A combined public key-based approach which is an offline authentication is proposed. A massive data authentication and cross-domain authentication problem of security is solved by this mechanism. This mechanism also provides security preservations for the data transmission and the sensor data. This mechanism provides the solution at the perception transmission and the application layer [18]. A framework called "context aware security" is suggested for the cyber physical system to help identify the behavior and study the surroundings. This new framework protective measures, sensing, and cyber are all interconnected. It's flexible for the CPS since it includes security controls, cryptography, and key agreement. The main objectives of this framework are integrity, authentication, confidentiality, and authorization [19].

7.6 Attacks on Cyber Physical Systems

These days the challenges in security and privacy have been a major cause of concern between customers and organizations because different cyberattacks are growing around the world. When the data is being transmitted through an open network between source and destination, the protection of sensitive data gets vulnerable. On this open network there is a probability of a cyberattack of extracting data or device disability for individual misuse [20]. The connected objects for example vehicles, smart homes, smart classes, and malls have been breached by hackers for gathering money and exposing the personal data. The arrangement of managing privacy and the safety of information and devices has become the center of focus for several companies. The different kinds of attacks are discussed further.

7.6.1 Network Attacks

The information could be revealed because of shortcomings in control and security. There are two types of network attacks that are passive attacks, which imply information is been observed, and active attacks, which imply data is being changed or altered [21].

If we do not have any plan for security, our information and network could be unsafe and prone to any type of attack. Those attacks normally exist on devices and networks. The frequent attacks on networks are information modification, network eavesdropping, ID spoofing (IP address spoofing), attacks based on passwords, denial of service, the middle man, application layer attack, the exploration attacks, the attacks on privacy, disastrous attacks, and so on.

7.6.2 Cryptographic Attacks

The process of ignoring the security of device/system after finding the vulnerability [22] in a cipher, security algorithm, operating systems, the cryptographic protocol, or management of the key system is called a cryptographic attack. The process is also termed crypto analysis. There are many classes of attacks that can be partitioned into crypto analysis. These vulnerabilities are susceptible to chosen plaintext, known plaintext, selected key, and crypto locker attacks.

7.6.3 Malicious Software

The softwares which are used for exposing cyber physical system infrastructures, breaching or stealing of data, and getting access controls of CPS are termed as malicious softwares. The main target of the malicious software is causing vulnerability to the host device. Because if it harms the host device it can breach the information of the devices connected through it. Rat-trapping, browser hijackers, different kinds of virus, worms, and phishing are the most used malwares by attackers these days [23].

7.6.4 Cyber Threats

It is also termed a malicious attack. It is also known as an intentional assault. With these assaults, hackers probe computer and network security flaws in order to compromise a targeted organization or person. A primary goal of cyberattacks is to cause damage to and severely limit the functionality of a system [24]. Our company faces a wide range of cyberattacks from the following key origins: EM leakage, harmonic distortion, and natural disasters (such as storms, flooding, device measurement failure, and node line faults) [25].

7.6.5 Cryptanalysis

The study of analyzing the hidden parameters of an information system is known as cryptanalysis [26]. It is the process of breaching the cryptographic system to achieve access to the data of encrypted messages, even when the cipher key is unknown. Cryptanalysis is also the study of the cryptosystem, cipher key, cipher text, and ciphers with the main goal of understanding how they actually work in order to improve the techniques for weakening them. In this, cryptanalysts try to decrypt the cipher text without the knowledge of plain text and the encryption key.

The main goal of cryptanalysis is to search for weaknesses in or to defeat the cryptographic algorithm. The result analyzed by the cryptanalyst is also helpful for the cryptographers to improve the flaws in the algorithm. Researchers also discovered some approaches of attack that fully crack an encryption algorithm, which may result in the cipher text which is to be encrypted by that particular algorithm and could be decrypted without knowing the encryption key as shown in Figure 7.3. When a flaw in the algorithm's implementations

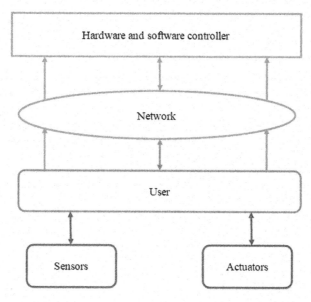

FIGURE 7.3
Basic block diagram of cryptanalysis.

is found, the number of keys that need to be tried to decipher the intended encrypted text is reduced.

Cryptanalysis attack types include:

1. Known-Plaintext Analysis (KPA): In this attack, the attacker decrypts cipher texts with the knowledge of the known partial plaintext.
2. Chosen Plaintext Analysis (CPA): In this method, the attacker uses cipher text that is similar to the selected plaintext by using the same algorithm technique.
3. Cipher text Only Analysis (COA): In this approach, the attacker has a large known collection of cipher text.
4. Man-in-the-Middle (MITM) Attack: In this attack, the attack is basically done when two users exchange a message via the communication channel that seems to be secure but actually is not.
5. Adaptive Chosen Plaintext Attack (ACPA): in this method, the attacker uses the cipher text and chosen plain text based on the previous encryption results.

7.7 Advantages and Disadvantages of Cyber Physical Systems

There are so many advantages of using CPS and so many disadvantages of CPS. So this chapter presents the ten advantages and six disadvantages of using cyber physical systems.

7.7.1 Advantages of Cyber Physical Systems

1. Cyber physical system is the quickest method to make sure of protection in actual global operations.
2. CPS makes sure of good effectiveness in different actual global operations.
3. There is an advancement of life standards for millions of people because of cyber physical systems.
4. Cyber physical system has the capability to conduct an optimistic revolution on a global level.
5. Cyber physical system can execute millions of calculations instantly.
6. Variability in cyber-physical operation of the system. The process of demonstrating a plan's repeatability is called proof-of-concept.
7. Cyber physical system is capable of developing and processing through new and untrustworthy circumstances.
8. Cyber physical system is capable to supply better performance with regard to feedback and automated re-design with a close interconnection of sensors and cyber architecture.
9. As a thread of cloud computing cyber physical system is capable of giving assets to users in terms of their needs.
10. Cyber physical system has the potential to supply more resources than wireless sensor networks and cloud computing on its own.

7.7.2 Disadvantages of Cyber Physical Systems

1. Cyber physical system gives less efficiency with regard to assets and potentiality. The unbalanced type of cryptography makes its instantaneous interaction vulnerable because of the extra time taken by encryption.

2. Due to its all-automated nature, it creates unemployment in the world.

3. Because of cyber physical system's real-time communication, it creates unpredictability.

4. In terms of cyber physical system forensics, it is vulnerable to many problems with regard to the deficiency of tools, expertise, and feedback against any non-forensic activities.

5. The efficacy of Intrusion Detection Systems is reduced by the fact that there are many distinct kinds of these systems, including sign-based, action-based, and peculiarity-based systems, which are often deployed in the context of Internet of Things subdomains.

6. By cyber physical system the devices could achieve their own consciousness. This can actually be a very big challenge in the coming future.

7.8 Applications of Cyber Physical Systems

1. **Green architectures**: Greenhouse effect is now one of the world's most pressing problems. Up to 72% of all produced power is used by the older facilities, leading to the warming of the atmosphere that, in turn, enhances the greenhouse gases. We want to achieve our goal of zero net energy consumption by using an integrated WSN, control mechanisms, and cognitive management.

2. **Smart grid**: This ecosystem's decision-making and management abilities will hinge on the quality of the data upon which they are based. Several conventional parts are used in cyber physical system by the smart grid. Smart grid is being used in four processes that follow: distribution, generation, consumer side, and also in transmission as well. The interconnection, networking, and working features within the electrical generation are controlled by smart grid. Cyber physical systems monitor and secure the smart grid's linked distribution and communication networks. Smart grid allows for two-way communication and management of the electricity grid by its customers [27].

3. **Medical cyber physical system**: The observation of well-being and medicines management of patients and gathering of diagnostic data is done by wireless sensor networks (WSN). A foundational requirement to high-credence medical CPS is given by the gathering of calculating and controlling contrivance to the important medical data transmitted [28].

4. Smart teaching could also be done by cyber physical system. The appropriate data about the physical environment is integrated by the cyber physical system in the smart learning environment, gives important and prompt assistance to students, teachers, and colleges, and ultimately transforms computed information into data

and findings. The Software Learning Environment (SLE) will ultimately revolutionize the way people study and work in colleges.

5. The cyber physical system can be used in flight implementations for example air test instrumentation, structure well-being observation, the landing of flight, the entertainment wireless compartment in the flight, and staff-pilot intercommunication etc.

6. In recent times, many civil engineers face the challenge of management of getting old of architectures like flyovers, buildings, dams, etc. A mechanical sensor, fiber optic sensors, and micro electrical sensors and wireless intercommunication automation give enormous assurance for accuracy and uninterrupted architectural observation.

7. Cyber physical system gives a path to upgrade the performance of control of traffic system [29]. An environment that lies in the natural geopolitical environment and artificial environment, for example, flyovers across the ocean or waterways, and huge underpass, high-threat sub-grade slope, city uplifted flyovers are assembled by road traffic control by cyber physical system, and also tremendous diversity of vehicles, humans, and different goods in the road environments [30].

8. **Humanoid robots**: Humanoid robots of this sort are employed to provide care for elderly people who live at home or in other locations where they may be easily accessed by the robots. These types of robots are used in underwater sea operations, muddy military operation, landmine detection, space operations, rainforest environments, and critical infrastructure problem-based task. They are used for personal use and they are also used in the agricultural field. These types of robots are especially helpful in emergency scenarios and violent confrontations when human lives are under jeopardy.

9. **Manufacturing**: The cyber physical system is also used in the manufacturing department for monitoring and controlling operations. Cyber physical system also improves the manufacturing process by distributing information between the business system, suppliers, customers, supply chain, and machines. For improving the traceability of the security goods, the cyber physical system does that work easily by doing a certain set of operations [31].

10. **Water distribution cyber physical systems**: Nowadays, water distribution systems have become more automated. Tanks, pumps, wells, pipes, and reservoirs are used for creating these types of automated water distribution systems. The sensors are used for detecting the overflow of water in tanks and pipes, and the algorithm is used for automated operations likes opening of valves, data acquisition, supervisory and monitoring control, and all the devices which lie in the network. They uncover the systems to possible attacks on the related software that has control of it. If the attacker can get the remote access of any component of the cyber physical system so they can also do damage like stealing data, cut water supplies, and damage equipment.

7.9 Conclusion

This chapter presents the security and privacy issues in cyber physical system. The literature reviews of eight papers have been done in this chapter which is related to cyber physical systems. A general architecture of the cyber physical system is also presented in this chapter along with its components. A simple framework for the cyber physical system used in transportation system is also proposed along with the description of the proposed framework. The attacks on cyber physical systems are also elaborated on in this chapter. The advantages and disadvantages are also presented in this chapter. The applications of cyber physical systems are also elaborated. The latest research topics and future research areas in cyber physical systems are also discussed. This chapter will help future researchers working in this area and related topics because this chapter contains all the security- and privacy-related aspects of cyber physical systems along with the proposed framework, architecture, pros and cons, and the applications.

References

1. R Baheti, H Gill. *Cyber Physical Systems*. The Impact of Control Technology, 2011.
2. GA Fink, TW Edgar, TR Rice, DG MacDonald, CE Crawford. Overview of security and privacy in cyber physical systems: Foundations, principles and applications. In *Security and Privacy in Cyber-Physical Systems* (pp. 1–23). Oak Ridge National University, Pacific Northwest National University USA, 2017.
3. J Shi, J Wan, H Yan, H Suo. *A Survey of Cyber Physical Systems*. IEEE, 2011.
4. J Giraldo, E Sarkar, AA Cardenas, M Maniatakos, M Kantarcioglu. *Security and Privacy in Cyber Physical Systems: A Survey of Surveys*. IEEE, August 2017.
5. A Humayed, J Lin, F Li, B Luo. *Cyber Physical Systems Security – A Survey*. IEEE, 17 January 2017.
6. EK Wang, Y Ye, X Xu, SM Yiu, LCK Hui, KP Chow. Security issues and challenges for cyber physical system. *2010 IEEE/ACM International Conference on Green Computing and Communications & 2010 IEEE/ACM International Conference on Cyber, Physical and Social Computing*, China, 2010.
7. F Pasqualetti, F Dörfler, F Bullo. Attack detection and identification in cyber physical systems. *IEEE Transactions on Automatic Control*, 58(11), November 2013.
8. R Rajkumar, I Lee, L Sha, J Stankovic. Cyber physical systems: The next computing revolution. *Design Automation Conference, ACM*, Anaheim, CA, USA, 2010.
9. Z Cai, X Zheng. A private and efficient mechanism for data uploading in smart cyber physical systems. *IEEE Transactions on Network Science and Engineering Department of Computer Science*, Georgia, State University, Atlanta, GA, USA, 2018.
10. MHR Muneeb Ul Hassan, J Chen. Differantial privacy techniques for cyber physical systems: A survey. 27 September, 2019.
11. Y Ashibani, QH Mahmoud. Cyber physical systems security: Analysis, challenges and solutions. Department of Electrical, Computer and Software Engineering, University of Ontario Institute of Technology, Oshawa, Elsevier, Canada, 12 April 2017.
12. R Alguliyev, Y Imamverdiyev, L Sukhostat. Cyber physical systems and their security issues. Elsevier, Institute of Information Technology, Azerbaijan National Academy of Sciences, 9A, B. Vahabzade Street, Baku AZ1141, Azerbaijan, April 2018.
13. G Yang, CM Rong, C Veigner, JT Wang, HB Cheng. Identity-based key agreement and encryption for wireless sensor networks. *The Journal of China Universities of Posts Telecommunication*, 6:54–60, 2006.
14. SN Premnath, ZJ Haas. Security and privacy in the internet-of things under time-and-budget-limited adversary model. *IEEE Wireless Communications Letters*, 4(3):277–80, 2015.
15. W Trappe, R Howard, RS Moore. Low-energy security: Limits and opportunities in the internet of things. *IEEE Security & Privacy*, 13(1):14–21, 2015.
16. J Wang, H Abid, S Lee, L Shu, F Xia. A secured health care application architecture for cyber-physical systems. *Control Engineering and Applied Informatics*, 6: 101–108, 2011.

17. MM Fouda, ZM Fadlullah, N Kato, R Lu, XS Shen. A lightweight message authentication scheme for smart grid communications. *IEEE Transactions on Smart Grid*, 2(4):675–685, 2012.
18. B Zhang, X Ma, Z Qin. Security architecture on the trusting internet of things. *Journal of Electronic Science and Technology*, 9(4):364–367, 2011.
19. EK Wang, Y Ye, X Xu, SM Yiu, LCK Hui, KP Chow. Security issues and challenges for cyber physical system. In *2010 IEEE/ACM Int'l Conference on Green Computing and Communications & Int'l Conference on Cyber, Physical and Social Computing* (pp. 733–738). 2010.
20. G Wu, J Su, J Chen. Optimal data injection attacks in cyber physical sytems. *IEEE Transactions on Cybernetics*, 48(12), December 2018.
21. F-J Wu, Y-F Kao, Y-C Tseng. From wireless sensor networks towards cyber physical systems. Elsevier, Department of Computer Science, National Chiao Tung University, Hsinchu, Taiwan, March 2011.
22. D Serpanos. *The Cyber Physical System Revolution*. IEEE, University of Patras, Greece, 2018.
23. A Singh, A Jain. *Study of Cyber-Attacks on Cyber Physical Systems*. Elsevier, 3rd International Conference on Advances in Internet of Things and Connected Technologies (ICIoTCT), Delhi, India, 2018.
24. M Krotofil, A Cárdenas, J Larsen, D Gollmann. *Vulnerabilties of Cyber Physical Systems to Stale Data Determining the Optimal Time to Launch Attack*. Elsevier, Department of Computer Science, USA, October 2014.
25. RM Góes, E Kang, R Kwong, S Lafortune. Stealthy deception attacks for cyber physical systems. *2017 IEEE 56th Annual Conference on Decision and Control (CDC)*, Melbourne, Australia, 12–15 December, 2017.
26. Y Xiao, Q Hao, D Yao. Neural cryptanalysis: Metrics, methodologies and applications. *IEEE Conference on Dependable and Secure Computing, Computer Science Virginia Tech Blacksburg*, 2019.
27. M Bhrugubanda. A review on applications of cyber physical systems. *IJISET - International journal of innovative science, engineering and technology*, 26: Telangana, India, June 2015.
28. IJISET - International journal of innovative science, engineering and technology, Vol. 2, Issue 6, Telangana, India, June 2015.
29. C Konstantinou, M Maniatakos, F Saqib, S Hu, J Plusquellic, Y Jin. Cyber physical sytems: A security perspective. *20th IEEE European Test Symposium (ETS)*, USA, 2015.
30. STW Xu, G Jizhen, Y Chen. The analysis of traffic control cyber physical systems. *13th COTA International Conference of Transportation Professionals (CICTP-2013)* (pp. 2487–2496).
31. X Guan, B Yang, C Chen, W Dai, Y Wang. A comprehensive overview of cyber physical systems: From perspective of feedback system. *IEEE/CAA Journal of Automatica Sinica*, 3(1), January 2016.

8

The Impact of Industry 4.0 Cyber Physical Systems on Industrial Development

Sanjay Kumar, Sahil Kansal, Kuldeep Singh Kaswan,
Vishnu Sharma, Javed Miya, and Naresh Kumar

CONTENTS

8.1 Introduction

An automobile engine programmer might have a price tag of over US$2 billion (one-third of a production line's costs are spent on setup and maintenance). As modern manufacturing systems technologies such as the Internet of Things (IoT) and intelligent goods are becoming more common, smart manufacturing systems can reduce costs while meeting demand. These systems are designed to be self-organizing, self-maintaining, controllable in real time, efficient, robust, and autonomous [1]. One of the terms established by several research groups worldwide to describe the evolving paradigm is Endeavour Industries 4.0, a German initiative to redefine manufacturing. A system composed of dispersed components and closely linked digital representations, commonly called "twins." These

DOI: 10.1201/9781003220664-8

systems are referred to in academic literature as cyber bodily systems. Digital modeling and the IoT can often bring the digital and physical worlds together and provide unique item identification in connection with materials, equipment, goods, and people [2]. It has been recognized that CPSs can meet clients' unique demands while also establishing and creating a creative production model in the health check, power, and communications sectors. By leveraging better defect detection to raise production yield and minimize waste, cyber physical production systems (CPPSs) improve product quality, production efficiency, and system flexibility. The availability of real-time, current, and historical system data facilitates the development of credible simulations for planning, production preparation, load complementary and flexible, decentralized labor reserve organization to meet fluctuating demands. As part of the "smart manufacturing" paradigm [3]. Industrialized and logistics systems should be updated capabilities to increase the visibility of goods and processes and optimize resource utilization and sustainability practices. At the local, national, and international levels, reliable assets and products are essential to creating sustainable supply chains. Radiodiffusion Television Ivoirienne (RTI), such as motionless ages racks cases, is widely old to collect stock and optimize the logistics process. The use of RTIs helps reduce solitary use wrapping and safeguard mechanisms from injury and ecological threats throughout transit to meet global sustainability goals. Packaging and RTIs are important in the supply chain because they allow component aggregation that are keen on a solitary body, which improves logistical competence. Still, they also serve as tools for aggregating components into a single entity, which improves logistical efficiency [4]. RTI monitoring and management systems have to be adaptable and changeable to suit changes in demand and diversity as part of "smart manufacturing." Only 33% of SMEs see the immediate benefits, and 80% believe their IT infrastructure is incapable or inadequate to enable such use. Regardless of their variety of payback, RTIs have downsides such as inefficient logistical procedures and conclusion hold up [5]. RTI is frequently absent, deleted enduringly, or injured (i.e., rendered worthless). Over 10% of RTIs need to be replaced every year, with another 10% requiring maintenance, with one example costing US$2 million. The direct costs of replacing and updating RTIs are straightforward to calculate, but a slew of secondary expenses have far-reaching implications for a company [6–8]. Indirect costs include missing components (both completed and in progress), late component entrance into manufacturing processes, and the provision of defective components. Everybody loses when factories stop producing and people stop spending money Additionally, determining the appropriate RTI navy dimension can be difficult; an enlarged fleet wastes resources, whereas an undersized fleet fails to satisfy demand necessitating the development of a system to aid in the monitoring and optimization of RTI use. Two such difficulties are extreme client investment time and harm caused by abuse. An ECHT of RTIs may be attributed to a number of causes, including incorrect demand forecasts, a lack of necessary facilities, a delay in initiating necessary arrangements, and a failure to fully appreciate the associated costs. About ten percent of RTI's fleet disappears every year due to theft. For effective RTI returns, companies often rely on personality coverage and kindness among stakeholders. To detect, treat, and remove damaged RTIs from circulation, companies must guarantee that RTI surveillance technology with techniques is inside put. RTIs' existing visibility leads to a lack of procedure possession, delaying problem resolution. The present is a necessity toward guarantee monitoring and tracking RTI efficiently is input with the configuration of contracts, incentives, fines, and preservation commitment inside and flanked by organizations [9]. This may aid in reducing arguments about who is accountable for replacing RTI casualties and whether or not the decisions made were acceptable. The RTI monitor system must gather and present information on

RTI input presentation indicators (KPIs) to overcome the constraints imposed by cyber physical production system (CPPS) standards and multi-stakeholders. RTI data must be precise plus obtained in "real-time" as of the provided chain to make suitable judgments. RTI fleets may demand regular real-time data collection, necessitating a low-cost solution. RFID's use to provide real-world instance data collection makes it possible to build a tracking system with little outlay of resources by facilitating Automatic Identification (AIDC) of assets. This paper investigates the tools necessary to design a sophisticated RTI monitoring system, with the goal of improving management and transparency around RTI implementation [10]. This chapter which was carried out in partnership with the automaker was used to improve CPPS capability with minor communications adjustments. RTI usage management (to eliminate bottlenecks and delays) and quality assurance for RTI utilization objectives, which are a significant element of their production processes for decreasing engine defects, were the two aims of the smart RTIs monitoring system. The previously provided research on smart RTI design and implementation is examined to help expand the prospect system, therefore causal to the corpse of proof supporting the acceptance of CPPS and its possible use in elegant RTI organization. The following is the chapter structure: The section "Literature review and motivation" reviews relevant literature and provides the rationale for this inquiry [11]. The unique system architecture is offered in section "Smart RTI system design," and the procedures for the evaluation case study are detailed in section "Research methodology." Section "Case study within an automobile factory" describes the implementation. In contrast, section "Conclusions and additional work" describes the outcomes and future work [12].

The requirement for suppleness and receptiveness among organizations, as delicate as a multifaceted and incessantly changing corporate surroundings and customer insist, are pushing knowledge advancement. Technology has long been recognized as a critical strategic instrument for assuring an organization's long-term success and effectiveness. Many companies are already using e-business technology to improve operational efficiency by streamlining their business processes and achieving convergence [13]. They're spending a lot of money on automation and robots right now so they can take advantage of Industry 4.0's cutting-edge innovations in the manufacturing sector. Industry 4.0 has been predominantly examined from production as an operational vision in recent literature and industry papers. On the other hand, Industry 4.0 and the technology that enables it can transform each feature of factory plus organizations, vastly enhancing executive disciplines such as providing sequence and logistics organization [14]. As a result of the continual rise in business system automation, companies are adapting the manufacturing 4.0 in-service model to another aspect of operations and the efficiency and productivity gains and quality enhancements that come with it. What makes this research stand out is its endeavor to provide a comprehensive and integrated shift toward implementing technological innovations connected to Industry 4.0 in the service of providing sequence organization [15]. Third, it presents an incorporated framework for illustrating how embracing Industry 4.0-related technology improvements could boost supply chain performance.

8.2 Literature Review

According to the authors, tags can be classified depending on their authority and communiqué capabilities. Since the event wave is a modulated multispectral signal from a passive ticket, it may be utilized as a transmission media. It must be reachable via antennas. A partially dormant label retains contact with its authoritative source via altered backscatter transmissions. Lively tags have inbuilt power, can broadcast signals, enhance location accuracy, and need electricity, making them more luxurious to put up and uphold. A mechanism for tracking active tags in the grocery supply chain was created to track RTI. Because of the variety of settings and the necessity to check hotness, with the requirement to execute all monitors "on product" relatively than on communications, active tags were chosen [16–18]. Since dynamic tag technology had a deeper comprehension of writing range, read/write capabilities, and the capacity to accept additional production orders, it was compatible with the cold sequence supply system. Although the authors claimed their method was cost-effective, they have not yet built a prototype or offered estimates of what they would cost to create. You may need to know when a creation enters and exits particular product traceability areas. This 462 may be done by placing an inert RFID tag on a crop and placing a reader in the places that need to be monitored. Technology is an industrial magazine that focuses on cutting-edge research and development in that field. Neal et al. However, while employing RFID tags, it's crucial to think about which tags to use and where they'll go on things because some materials might cause interference. This may be avoided by employing proper identifiers, such as the RFID bolt described, which is used to track crankshafts throughout milling and assembly. However, the crankshafts underwent an extra balancing operation when the bolts were removed from the items. An active new upgraded tag system was used [19]. The technology was developed modestly in a real-world context with an automobile industry partner. Innovative technology provides better granularity to RTI and its state. The vigorous improved tag scheme necessitated the supply of a continual power source, which had consequences for maintenance and charging. Furthermore, low-cost deployment of such a system is no longer possible due to the increasing difficulty and linked expenses of the different mechanisms. Passive RFID is a comparatively reasonable approach for establishing a unique identifier inside constructing elegant substance and capturing data from physical processes for use in the digital world. An inert tag can also be worked to sense hotness and dampness to maximize the data collected. On the other hand, these tags will only work if there is RF power present, which may necessitate the installation of supplementary communications in areas where sensor records are necessary. A half-inert or lively tag intended for footage antenna data may be more cost-efficient [20]. The labels used and the system's setup impacts a traceability system's cost. Antennae in communication with middleware vary depending on the range of organization and storage space.

The widespread use of RFID and sensors across a system has caused a meteoric increase in information creation and has performed a thorough, collaborative analysis of a massive dataset including RFID ticketing data. In yet another approach to RFID data management, an RFID Cuboids model is employed to organize data based on event series and manufacturing rationale. The data might then be used to trace down where the products went and compile statistics to aid in decision-making. Solutions have been shown to reduce data volume effectively. Increased usage of tagging and sensors across the system results in more data and increases the danger of misreads. As a result, it is attractive to have an information organization system to organize and sieve pertinent data. However, neither technique considered what would happen if a tag or piece of information was not read in a real-world scenario; it's feasible that a location might vanish.

Although the research did not clearly show system efficacy or possible obstacles to system adoption in the industry, they must be analyzed to identify the recommended real-world benefits. Many research studies looked at analytically; for example, Kang and Lee suggest a generic RFID monitoring system evaluated using an algorithm in a "typical" instance. To reduce human operator cargo handling mistakes. Traceability improved when additional degrees of track and classification was added. Still, the price was more significant than, according to the findings, intermediate implementation findings would benefit both minimizing cargo loss and lowering implementation costs. Even though an industry partner was brought in to document current operations and offer input on simulation findings, demonstrate a variety of RFID-based tracking and traceability approaches for improving industrial processes. For example, a system for tracking goods and methods in intra-company logistics and production was established [21]. Corporeal pattern, request repair, and communal communication levels were part of the three-layer framework. As essential services as part of this system, (i) real-time production and transportation data were collected, (ii) real-time data analysis and monitoring, and (iii) decision support measures, such as capability evaluation and task integration, were selected. An arithmetical replica was shaped to recognize the best manufacturing and shipping alternatives based on time and cost. An imaginary case study was provided for the system, but it was never used. It is used in the real world and relies on flawless system operation. They also created an RFID-enabled tracking system for modeling production flow, with four RFID tracking application scenarios to choose from including fixed and mobile "antenna-reader" and "detection space control mode" combinations. Some of the request scenarios that have been found are as follows: (i) I installed an RFID antennae/reader at a location where tagged objects pass through. (ii) RFID antennae/readers that are fixed in a space where tagged things pass; (iii) RFID antennae/readers that are fixed in a space where tagged items pass; (iv) RFID antennae/readers that are fixed in a space where tagged items pass; (v) a reader installed on a mobile vehicle that tracks item movement (vi) gateway antennae that track people and assets entering and leaving, and (vii) a mobile scanner with pre-programmed tag locations. Many of the concerns that may arise during the deployment of a real-world system, such as data privacy restrictions, distrust and misunderstanding of the process, and changing demands, were not present during the case study examination, which took place in a sterile laboratory setting [22]. We suggest employing an active RFID tag for greater location monitoring of automated guide vehicles based on real-world, potential concerns seen during deployment of more autonomous systems and CPPSs that use an automated vehicle. A mathematical model is used to show and verify the results, but it is not experienced in a genuine earth case. The focus of the text review was mostly on CPS and RFID models and laboratory and theoretical container study, with real-world implementation lessons being overlooked. Despite a few examination beds in areas like an elegant energy grid and smart city, few academics have focused on physical RTI deployment in the industrial sphere. The authors found no verified architecture in the literature to hold the creation of the RTI tracking system, encompassing software growth, addition inside CPPS, and the Industry 4.0 example, as well as deployment. It is critical to evaluate how these tracking systems labor in the genuine earth to ensure their applicability and implementation in business. A real-world innovative system implementation study also aids in identifying and resolving the issue that might obstruct extensive use of this technology in the industry, limiting the realization of potential CPPS benefits. This publication contains the findings of this inquiry [23] (Figure 8.1).

We suggest employing an active RFID tag for greater location monitoring of automated guide vehicles based on real-world, potential concerns seen during deployment of more autonomous systems and CPPSs that use an automated vehicle [24]. The architecture described here and the lessons learned may be applied to a broader range of domains where current infrastructure must be modified to intelligent capabilities.

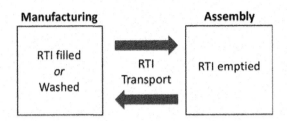

FIGURE 8.1
SOA and RAMI4.0 are used in the CPS RTI monitoring architecture.

8.2.1 Research Methodology

The research technique is based on the recommendations of Areola and Baines, and it employs an adapted human-centered plan method that is well-suited to the construction of CPPSs. The strategy was based on a published case study that was used to create, implement, and assess the sophisticated RTI program. Stakeholders need admission to pertinent information to improve human decision-making and choose the best course of action for achieving stated goals. The approach used the information knowledge Wisdom dominance hierarchy, which was introduced in [25], to guarantee that the scheme is built to facilitate excellent decision creation. Genomic sequences have been adjusted to give connotation and context, and information a collection of facts and, in turn, comprehension, estimate, and ability have been used to enhance decision manufacturing.

8.2.2 A Case Study Is about Automobile Manufacturing

Within their UK production and assembly factory, a large automobile manufacturer embarked on expanding an elegant RTI system. A competent RTI organization is needed to track RTI in a congested ring provide chain involving a crankshaft, camshaft, cylinder head manufacturing plant, and a training meeting plant [26]. Every RTI comprises a weld strengthen border and a set of artificial separator trays that ensure component placement and prevent breakage.

Step 1: Recognizing Commerce Requirements and Process

Interviewing with identifying stakeholder's older organization, operation management, electrical and automatic repair team, and line-side workers were recognized as key stakeholders. Procedure comments, interviews, and interactions with key stakeholders were conducted to determine the standard operating procedure (SOP) for implementing RTIs. Logistics are on the line (Figure 8.2).

Activities are scheduled, monitored, and organized by coordinating all material handling. Management ensures that material moves smoothly through the facility. The function of the shop floor operator is to carry out the material handling activity physically. This involves communicating with RTIs, separators, and components throughout the two sites to guarantee that the proper materials and components are continuously provided to the manufacturing and assembly lines. Business principles include things like electrical wiring, software, and hardware, and it is the responsibility of the authority to ensure that any plan implemented on the supermarket floor does not violate these standards [27]. The mechanical and electrical services crew is responsible for ensuring proper setup and commission of supermarket ground gear after the PTME team has approved it. The team

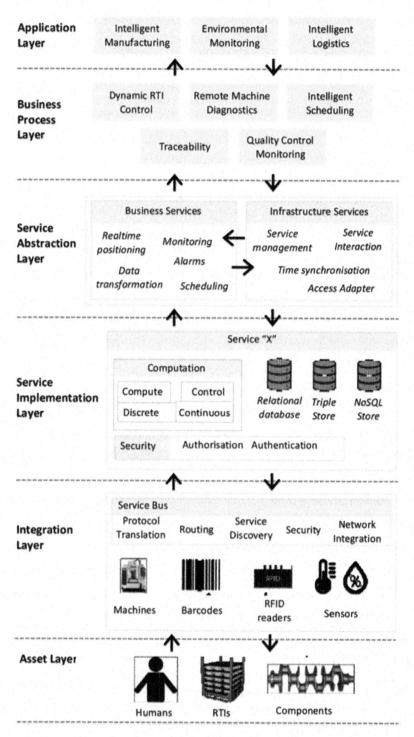

FIGURE 8.2
Framework of requirement process.

oversees all manufacturing processes and ensures that the plant runs smoothly and meets the corporation's distinct presentation criteria. All stakeholders necessitate the capacity to monitor and review presentation indicators intended for every one sector of the amenities, counting the monitoring and management of RTI, to improve the company's efficiency. The logistics team, which includes the store's management and front-line employees, is the most important group of stakeholders for an RTI to track and manage.

8.3 Process Identification and Mapping

The experimental processes were codified using the Business Process Modeling and Notation (BPMN) paradigm for stakeholder verification. A simplified BPMN diagram depicting the primary functions in connection to the flow of RTIs with operations components eliminated at the manufacturing or meeting stages at the finish of a part manufacture line, RTIs are shipped empty. Where they are loaded with WIP or completed parts, they're kept in the manufacturing marketplace when they're full. Due to a lack of stock at the meeting facility, a demand mechanism will be dispatched to the urbanized marketplace in search of the necessary RTI, from whence it will be transported by truck. The cargo specifics are sent to the lorry driver once it has been completed [27]. The RTIs are brought to the assembly factory in their whole by the driver. Complete RTIs are delivered to the assembly marketplace located inside the assembly plant. Empty RTI waits in the assembly marketplace to be returned to the manufacturing facility. On the production line, unfulfilled RTIs are sent back to the meeting market. In order to keep up with the needs of the train meeting plant, which attempts to construct one engine every 30 seconds, this method is kept going forever. Placing a separator on an RTI, filling it with components, moving it to the assembly line, discharging it, and returning it to be refilled constitutes a single cycle. A 12-hour assembly schedule was proposed to provide convenient transit between manufacturing and assembly, lowering the danger of surface corrosion. During BP2.8, empty RTIs are visually examined to check if they need to be cleaned after being unloaded from the assembly factory. Separators can become polluted over time due to their proximity to manufacturing activities such as drilling, cutting, and milling and interactions with other Work in Progress (WIP) or transportation components [28]. When separators get unclean, there is a danger of contamination of the final parts. Engine failure or early deterioration can occur due to a harmful component and a drop in expected performance. To maintain track of this issue, and as part of an RTI management CPS, intelligent separator and RTI cleaning methods must be implemented. RTIs and separators are unidentifiable, and the washing machine lacks AIDC technology. As a result, the washing process is currently unaudited, making any technique for increasing cleaning frequency. Long client hold time deprived place recognition, and reduction are all difficulties that the company faces, with RTI fleet replacements costing up to RTI. These concerns are similar to those raised in the literature review. Visual inspections of RTI racks and separators are performed, and basic training is offered to aid in recognizing probable pollutants and estimating the contamination threshold for cleaning an RTI plus its partition [29]. There are currently two methods for cleaning unclean separators and RTIs: (i) an automated washer to jet clean and atmosphere dry the mechanism at 42 _C, and (ii) a physical plane cleaning technique intended for RTI frame that uses a combination of manufacturing cleaner and irrigates, follows by compulsory convection ventilation. The washing method now follows

TABLE 8.1

Users and Their Responsibilities Are Identified

Stakeholders	Domain	Role/responsibilities
Logistics management on the production line	Logistics	She supervises the logistical management and material scheduling throughout the industrial plants.
Logistics on the line	Logistics	I must physically handle RTIs, separators, and components.
Mechanical and electrical Services	Installation and upkeep of hardware	Responsible for facility maintenance and the installation of additional manufacturing cells on occasion.

a predetermined schedule that distributes time evenly among all separator types. RTI's condition or demand does not appear to be used to prioritize laundry. Since the examination approach is physical and therefore based on human insight, this work gives a wide range of quality (Table 8.1).

8.3.1 Decisions about Operations

The bulk of RTI-related choices need data about the RTI population's present and historical statuses. There were five major operational decisions to be made: (D1) Does the RTI or Separator need to be washed. Where may empty or full RTIs be found? Do RTI processes need to be changed? The Analytic Hierarchy Process (AHP) was used to reduce the judgments necessary to achieve RTI observability and traceability to their underlying knowledge and information components. AHP is a technique for analyzing the priority of every one supplies by breaking down a far above the ground level aim, such as RTI monitoring and control is broken down into the layer of different requirements, such as choices data, in order, in addition to statistics. To evaluate which decisions deserve support, the value of each decision is weighed against the importance of all other choices [29]. This is done on a scale of one to nine, with one denoting equal importance and nine indicating the most significant possible importance above the rest. All decisions that need to be supported, as defined by the stakeholders, are compared. The main concern level of any option is strong minded by adding all of the association principles in its row and separating by the figure of all rows, as long as the primary anxiety value. The precedence principles in Table 8.1 illustrate that the judgments are in order.

8.3.2 Organizing Needs by Priority

Based on the DIKW hierarchy, the Sankey diagram depicts the relations flanked by decision, information, and data supplies and their relative importance. The level is based on the Quality Function Deployment approach, which employs the ranges to priorities demands (high correlation). The top three criteria identified throughout this technique were an acquaintance of the dispensation pathways (RTI position (K3) (27%) and RTI sharing (K2)). The figure of RTI cycles and the time it takes to transfer between facilities (I2) are two factors to consider [30]. To hold up the in sequence, acquaintance, and judgment, 13 information types were chosen to fulfill the senior height needs. RTI IDs are regarded as high-priority data categories, as with time stamps (12%) and geographical identifiers (12%).

8.3.3 Nonpractical Specifications

Scalability guarantees that the arrangement can be scaled to convene any nicotinamide adenine dinucleotide insists by participatory processes message proper procedure and scheme reserve complement capabilities – additional RTI, facilities, and user, for example. Stakeholders have specified additional non-functional requirements, such as flexibility, toward a deal by an unexpected change in RTI route and commerce process. (iii) Modularity allows for the incorporation of future technology developments and unique RTI types, such as on-plank RTI sensors. (iv) Sanctuary guarantees any CPS will manage and control its resources. Third-party attacks such as "man-in-the-middle," denial of service, and insertion of false data can lead to unauthorized disclosure [31]. Support for these non-functional criteria is essential for some, but there are several stakeholder requirements that academic research hasn't previously identified when considering future technologies and system implications. The use case revealed the following necessities: Existing conventional skillets that may be enhanced utilizing industrial hardware/software architectures are used or enabled inside a CPS. The organization's supplies are defined, biased, and discussed by the stakeholders. The agreed-upon materials for the classy RTI system are as follows: The following factors are included into my calculations: (i) travel time, (ii) proximity to amenities, (iii) compliance with standards, (iv) ease of completion, (v) data security, and (vi) system viability.

8.4 Layers of Overhaul Company and Application

The idea of the secure online space was to facilitate customer dialogue and decision making in light of program data. For data protection, we employed both a protected registry and a classification basis. A set of interfaces was created within the GUI in compliance with human-computer interaction principles. The interfaces were created using a pattern arrangement and was iteratively experienced for usability. A navigation panel on the user interface allows users to access various pages such as reports, system status, administration, and system administrator contact information (denoted A). Additional system information is shown based on the choices of system stakeholders. The navigation panel offers stakeholders other details about the RTI, such as customizable separators [32]. A number of data analytics base GUIs should be developed to meet the wants of defining RTI navy size and commercial procedure devotion or version. As decided by the automobile manufacturer, the extra page permits stakeholders to look further into the data and pinpoint any issues with the RTI submission process.

In a similar vein, the histogram of time flanked by wash and the figure of a cycle a partition passes from surface to side before a life form is washed are exhibited to recognize the supremacy of drive effectiveness, RTI, and clean partitioning devotion. In each case, the stakeholders requested that a count and increasing incidence be displayed; the system engineers implemented these parameters in the administration panel. Any display can be used depending on stakeholder preferences owing to the supple method the GUI was designed. Using the data provided to the excellence boss, it is likely to recognize RTI that were not following the wash procedure and ensure that the marked things were washed before the subsequent use.

8.4.1 Analysis of a Case Study

Observations regarding the company's operations in terms of RTI flow and the difficulties that must be solved may be made due to the manufacturing use trial. Through the manufacturing usage trial, insights on the company's operations in terms of RTI flow and the challenges that must be addressed may be gained. Metrics like (i) time spent sleeping in a room with amenities, (ii) time spent putting together a mechanism, and (iii) time spent caring about the cleanliness of an activity have all been used to bring attention to critical issues.

8.4.2 The Amount of Time Exhausted in and out of Institutions

The observations below demonstrate how many instances an RTI spends in a different state, such as inside ability and traveling flanked by amenities. RTI uses fewer than 15 hours at the developed plant 50% of the time, according to statistics, while 21.3 percent of RTIs spend 80 hours or more in the facility. Because most of this time is spent idle, this indicates either an overcapacity or inadequate utilization of RTIs. RTIs stayed for fewer hours at the meeting facility, which had less storage room in its market, with 50% staying for less than 5 hours and merely 14.8% remaining for more than 50 hours.

When it comes to the occasion, it took to transport RTIs flanked by amenities 79.4% of RTIs to be delivered within 5 hours or fewer when traveling from the production site to the meeting factory, and 65.4% of RTIs were delivered in less than 5 hours on the go back support. This demonstrates a reduction in the importance of returning RTIs, as evidenced by the text. On the other hand, traveling to the production location took more than 20 hours, while RTIs took 12.9% longer to reach the assembly factory [32]. Based on this data, it's evident that there's potential for improvement in RTI implementation and management inside the hospital. Another issue addressed in the literature is RTIs that take excessive time to transfer because they are misplaced or placed in the wrong locations.

8.5 It's Time to Put the Pieces Together

The CPS experiment in place is a creative way to ensure that this condition is being met. The time it takes for an RTI to leave the manufacturer, travel to the assembly site, and be put together may be calculated as 53.2. Over the course of the 6-month trial, researchers discovered that % of RTIs required 10 hours or less of construction. However, 36.2% of RTIs were short of the required 12 hours [33]. The data would only be provided through if the CPS was not deployed, components and their logistical operations would be monitored manually. With this CPS implementation, the time-to-assembly measure may be monitored in real-time if the user has the requisite expertise in creating new changes and keeping tabs on their possessions

8.5.1 Devotion to the Wash Process

Every 10 cycles, at the very least, the separation has to be cleaned. The variety of cycles an RTI has accomplished is directly related to the number of processes it has completed, which is expressed as the number of conversions it has made across facilities. Collectors were only cleaned thrice throughout the six-month period. The first two occur after just one series, and the third after 20 cycles. This shows a problem with the company's quality control, which must be addressed [33]. As a result of the poor performance, the RFID integration in the washing machine was examined to make sure it was working correctly. No irregularities were established, implying that the clean devotion dimension was correct [33]. Because of CPS's continuous monitoring capabilities, the company may create planned commerce procedure enhancements, in addition to quickly assessing their repercussions.

8.5.2 Capturing Business Processes and Flexibility

The right business processes must be documented to use the RTI state progressions, locations, and wash adherence in a CPS model. When specific business procedures are not followed, i.e., erroneous processing routes and failure to detect the state of an RTI or separator, the state determination services cannot accurately determine the form of an RTI or partition to comply with established requirements. As a result, the system's CPS model is inaccurate, and any decisions based on it are most likely incorrect. This confronts puts force on the organization to identify correctly, implement, and regulate RTI management business processes and the CPS toward being flexible and recognizing what time procedure deviations happen [34]. The RTI condition will power CPS repair could be expanded to give stakeholders by recognized trend inside the dispensation route, allowing them to choose whether they ought to be included keen on the CPS as different RTI state or treated as wrong RTI dispensation, with stakeholders being notified if they carry on to happen.

8.5.3 Manufacture Informed Decisions

The CPS can be implemented in a limited way, but it might be enhanced to provide extra CPS base behavior and clever choice creation. Actual occasion decision on whether RTI cleaning is necessary might be introduced to aid stakeholders in what time is moving bare RTIs keen on the production plant. To evaluate if the RTI in transit should be transported to the wash area, the CPS can maintain track of the present RTI ease of use, a figure of a custom cycle, and wash mechanism utilization [34]. A CPS-equipped visual indication, such as a show or pointer glow, might tell the operator of their options. The data collected by the CPS could be utilized to create simulation and optimization services to help with automated RTI and material handling scheduling. This could assist a corporation in enhancing RTI usage by reducing inactive era and raising go back tax, resultant in lesser RTI navy size and cheaper expenses.

8.5.4 Mechanisms Are Being Integrated into the CPS

Only the RTIs and separators were included in the CPS for the container learn completion. RFID technology is being used to combine separate components that are being transferred. Once connected, the CPS would exist cleverly to give a clear genuine occasion view of part mobility inside the amenities, improve traceability and visibility of unique components, and better understand the time-to-assembly metric [35]. With this technology, the demand for trustworthy benefits toward CPS addition performance, i.e., using little tags that may be fastened to metal components during production, is difficult. Alternatively, the advent of intelligent containers, in which RTIs may identify the laden mechanism and interface with the CPS to give additional data, such as creation experience, could facilitate more excellent tracking [36]. Although the scheme was successfully installed with the findings being old for choice support, the system's more comprehensive application is limited due to the necessity to build a more robust system for extreme situations.

8.6 Conclusions and Further Work

This chapter discusses the study and creation of a CPS to improve RTI monitoring and management in the automobile industry. Rather than focusing on hypothetical, replicated, or trim level laboratory difficulties, this investigation has integrated a comprehensive CPS keen on entirely operating real-world car surroundings, anywhere RTIs are the procedure on top of a nonstop foundation. The CPS provided total visibility of RTI plus their associated separators to convene the wants of every significant stakeholder inside the area. As a result of this operation, empirical RTI usage data has been collected, allowing the company to examine the present condition of RTI procedures plus excellence standards. Several key RTI indicators (such as separator cycle, clean occurrence, an occasion to a meeting, and logistical procedure delay) can be compiled and made available in a secure GUI for stakeholder monitoring and decision-making. A proper CPS intend procedure was followed as a fraction of this completion, resulting in various information, knowledge, and choice needs that aid stakeholders in the RTIs and benefits organization areas. Constructing the CPS is likely to find difficulties linked to their implementations inside the room. Among the problems are the additions of genuine-earth assets with the digital globe, the documentation of commerce process, and the responsiveness of a CPS to procedure alteration. More study is needed into the growth of innovative choice-making services that can manage the run of RTIs and the option of integrating mechanisms inside the CPS. These improvements can be made by utilizing accessible RFID communications or constructing clever RTIs that use embedded sensors to detect and track components' experiences. More research is needed to assess the viability of such a scheme completion technique on a broader variety of demanding settings, such as those with a bit of communications, dampness, hotness, or powder.

References

1. Colombo, A. W., Jammes, F., Smit, H., Harrison, R., Lastra, J. L. M., & Delamer, I. M. (2005). Service-oriented architectures for collaborative automation. *31st Annu Conf IEEE Ind Electron Società, 2005. IECON 2005*, pp. 2649–2654.
2. Lee, J., Lapira, E., Yang, S., & Kao, A. (2013). Predictive manufacturing system—Trends of next-generation production systems. *IFAC Proc*, 46(7), 150–156. http://doi.org/10.3182/20130522-3-BR-4036.00107.
3. García-Arca, J., González-Portela Garrido, A. T., & Prado-Prado, J. C. (2016). "Pack- aging logistics" for improving performance in supply chains: The role of meta-standards implementation. *Producao*, 26, 261–272. http://doi.org/10.1590/0103-6513.184114.
4. Chaâri, R., Ellouze, F., Koubâa, A., Qureshi, B., Pereira, N., Youssef, H., & Tovar, E. (2016). Cyber-physical systems clouds: A survey. *Comput Netw*, 108, 260–278. https://doi.org/10.1016/j.comnet.2016.08.017.

5. Monostori, L., Kádár, B., Bauernhansl, T., Kondoh, S., Kumara, S., Reinhart, G., Sauer, O., Schuh, G., Sihn, W., & Ueda, K. (2016). Cyber-physical systems in manufacturing. *CIRP Ann Manuftechnol*, 65(2), 621–641. http://doi.org/10.1016/j.cirp.2016.06.005.

6. Lee, J., Bagheri, B., & Kao, H. A. (2015). A cyber-physical systems architecture for industry 4.0-based manufacturing systems. *Manuf Lett*, 3, 18–23. https://doi.org/10.1016/j.mfglet.2014.12 .001.

7. Barbosa, J., Leitão, P., Trentesaux, D., Colombo, A. W., & Karnouskos, S. (2016). *Cross Benefits from Cyber-Physical Systems and Intelligent Products for Future Smart Industries Cross Benefits from Cyber-Physical Systems and Intelligent Products for Future Smart Industries.* IEEE, pp. 504–509. http://doi.org/10.1109/INDIN.2016.7819214.

8. Leitao, P., Rodrigues, N., Turrin, C., & Pagani, A. (2015). Multiagent system integrating process and quality control in a factory producing laundry washing machines. *IEEE Trans Ind Inform*, 11(4), 879–886. http://doi.org/10.1109/TII.2015.2431232.

9. Li, D., Tang, H., Wang, S., & Liu, C. (2017). A big data-enabled load-balancing control for smart manufacturing of industry 4.0. *Clust Comput*, 20(2), 1–10. https://doi.org/10.1007/s10586-017 -0852-1.

10. Ardanza, A., Moreno, A., Segura, Á., de la Cruz, M., & Aguinaga, D. (2019). Sustainable and flexible industrial human-machine interfaces to support adaptable applications in the industry 4.0 paradigm. *Int J Prod Res*, 57(12), 4045–4059. https://doi.org/10.1080/00207543.2019.1572932.

11. Vega, D., & Roussat, C. (2015). Humanitarian logistics: The role of logistics service providers. *Int J Phys Distriblogistmanag*, 45(4), 352–375. http://doi.org/10.1108/IJPDLM-12-2014-0309.

12. Govindan, K., Soleimani, H., & Kannan, D. (2015). Reverse logistics and closed-loop supply chain: A comprehensive review to explore the future. *Eur J Oper Res*, 240(3), 603–626. http://doi .org/10.1016/j.ejor.2014.07.012.

13. Glock, C. H. (2016). Decision support models for managing returnable transport items in supply chains: A systematic literature review. *Int J Prod Econ*, 183, 561–569. http://doi.org/10.1016/j .ijpe.2016.02.015.

14. Wang, X., Ong, S. K., & Nee, A. Y. C. (2018). A comprehensive survey of ubiquitous manufacturing research. *Int J Prod Res*, 56(1–2), 604–628. http://doi.org/10.1080/00207543.2017.1413259.

15. Leitão, P., Rodrigues, N., Barbosa, J., Turrin, C., & Pagani, A. (2015). Intelligent products: The grace experience. *Control Eng Pract*, 42, 95–105. https://doi.org/10.1016/j.conengprac.2015.05 .001.

16. Segura Velandia, D. M., Kaur, N., Whittow, W. G., Conway, P. P., & West, A. A. (2016). Towards industrial internet of things: Crankshaft monitoring, traceability and tracking using RFID. *Robot Comput Integr Manuf*, 41, 66–77. http://doi.org/10.1016/j.rcim.2016.02.004.

17. Neal, A. D., Sharpe, R. G., Conway, P. P., & West, A. A. (2019). smaRTI—A cyber-physical intelligent container for industry 4.0 manufacturing. *J Manuf Syst*, 52, 63–75. http://doi.org/10.1016 /j.jmsy.2019.04.011.

18. Ding, K., Jiang, P., & Su, S. (2018). RFID-enabled social manufacturing system for inter-enterprise monitoring and dispatching integrated production and transportation tasks. *Robot Comput Integr Manuf*, 49, 120–133. https://doi.org/10.1016/j.rcim.2017.06.009.

19. Seshia, S. A., Hu, S., Li, W., & Zhu, Q. (2016). Design automation of cyber-physical systems: Challenges, advances, and opportunities. *IEEE Trans Comput des Integr Circuits Syst*, 1. http:// doi.org/10.1109/TCAD.2016.2633961.

20. Neal, A. D., Sharpe, R. G., van Lopik, K., Tribe, J., Goodall, P., Lugo, H., & West, A. A. (2021). The potential of industry 4.0 cyber-physical system to improve quality assurance: An automotive case study for wash monitoring of returnable transit items. *CIRP J Manuf Sci Technol*, 32, 461–475.

21. Alfian, G., Syafrudin, M., Farooq, U., Ma'arif, M. R., Syaekhoni, M. A., Fitriyani, N. L., & Rhee, J. (2020). Improving the efficiency of RFID-based traceability system for perishable food by utilising IoT sensors and machine learning model. *Food Control*, 110, 107016.

22. Kumar, S., Gupta, G., & Singh, K. R. (2015). 5G: Revolution of future communication technology. In *2015 International Conference on Green Computing and Internet of Things (ICGCIoT)*. IEEE, pp. 143–147.

23. Walter Colombo, A., Karnouskos, S., &Hanisch, C. (2021). Engineering human-focused industrial cyber-physical systems in industry 4.0 context. *Philos Trans R Soc Lond A*, 379(2207), 20200366.

24. Kravets, A. G. (2020). *Cyber-Physical Systems: Industry 4.0 Challenges*. A. A. Bolshakov & M. V. Shcherbakov (Eds.). Springer.

25. Pivoto, D. G., de Almeida, L. F., da Rosa Righi, R., Rodrigues, J. J., Lugli, A. B., &Alberti, A. M. (2021). Cyber-physical systems architectures for the industrial internet of things applications in industry 4.0: A literature review. *J Manuf Syst*, 58, 176–192.

26. Colombo, A. W., Veltink, G. J., Roa, J., & Caliusco, M. L. (2020, June). Learning industrial cyber-physical systems and industry 4.0-compliant solutions. In *2020 IEEE Conference on Industrial Cyberphysical Systems (ICPS)* (Vol. 1, pp. 384–390). IEEE.

27. Singh, H. (2021). Big data, industry 4.0 and cyber-physical systems integration: A brilliant industry context. *Mater Today Proc*, 46, 157–162.

28. Schiele, H., & Torn, R. J. (2020). Cyber-physical systems with autonomous machine-to-machine communication: Industry 4.0 and its particular potential for purchasing and supply management. *Int J Procurement Manag*, 13(4), 507–530.

29. Karnouskos, S., Leitao, P., Ribeiro, L., & Colombo, A. W. (2020). Industrial agents as a critical enabler for realising industrial cyber-physical systems: Multiagent systems entering industry 4.0. *IEEE Ind Electron Mag*, 14(3), 18–32.

30. Sinha, D., & Roy, R. (2020). Reviewing cyber-physical system as a part of the smart factory in industry 4.0. *IEEE Eng Manag Rev*, 48(2), 103–117.

31. Lozano, C. V., & Vijayan, K. K. (2020). Literature review on cyber-physical systems design. *Procedia Manuf*, 45, 295–300.

32. Kizim, A. V., &Kravets, A. G. (2020). Anepistemological approach to intelligent decision-making support in industrial cyber-physical systems. In *Cyber-Physical Systems: Industry 4.0 Challenges* (pp. 167–183). Springer, Cham.

33. Radanliev, P., De Roure, D., Van Kleek, M., Santos, O., & Ani, U. (2021). Artificial intelligence in cyber-physical systems. *AI Soc*, 36(3), 783–796.

34. Broo, D. G., Boman, U., & Törngren, M. (2021). Cyber-physical systems research and education in 2030: Scenarios and strategies. *J Ind Inf Integr*, 21, 100192.

35. Morella, P., Lambán, M. P., Royo, J. A., & Sánchez, J. C. (2021). The importance of implementing cyber-physical systems to acquire real-time data and indicators. *J*, 4(2), 147–153.

36. Sharma, R., Parhi, S., & Shishodia, A. (2021). Industry 4.0 applications in agriculture: Cyber-physical agricultural systems (CPAs). In *Advances in Mechanical Engineering* (pp. 807–813). Springer, Singapore.

9

Security and Privacy Aspect of Cyber Physical Systems

Umesh Kumar Singh[1], Abhishek Sharma[2], Suyash Kumar Singh[3],
Pragya Singh Tomar[4], Keerti Dixit[5], and Kamal Upreti[6]

CONTENTS

9.1 Introduction

Systems that integrate the physical and computational domains are known as cyber physical systems. Digital chips, software products, sensors, and actuators are among the diverse interacting components in CPS. As a result, the CPS ecology differs from and is more advanced than conventional settings. It is principally true because CPS is set up to become accustomed to its approach to the existing real-world environment in accordance to the observed condition [1]. CPSs are comparable to IoT systems; however, they feature greater physical and computational synchronization [2, 3].

CPSs interact with consumers, physical surroundings, and a range of hardware- and software-based systems. This includes the integration, interoperability, monitoring, and control of CPS components. CPS has a sequence of i/p and o/p connected to cooperating components, unlike stand-alone devices. Furthermore, CPS deployments are not restricted to a particular field; rather, they are found in nearly every field [4].

A CPS is defined through a wide range of organized technologies and a variable size across such structures [5]. These technologies, among other things, offer better personalization of health care, traffic control, finance, and the smart grid. Computing devices, embedded systems, sensors, control units, and other devices that perform various duties can be found in a CPS. A single CPS, for example, a room temperature surveillance and adjustment system might usually comprise a few sensor and actuator nodes. A CPS, on the other hand, can grow into a massive network of diversified and unstructured decentralized subsystems for example, which may do diverse independent jobs on a solar energy plant [6].

DOI: 10.1201/9781003220664-9

FIGURE 9.1
Cyber physical systems (CPSs).

Biometric technologies, IoTs, smart cities/industries/healthcare/agriculture/vehicles/ buildings and infrastructures, wearable technology, mobile communication, defense organizations, and climatology are just a few of the areas where CPSs are being developed under the guise of technical methodologies shown in Figure 9.1

The fast expansion of CPS implementations has resulted in a slew of security and confidentiality issues. Information security, as per ISO/IEC 27001: 2013, is the protection of information's CIA properties. CPSs have the flexibility to cope with this complication as well as changes in system scale. The volume and diversity of deployed components dictate the complexity of a CPS in most situations. Furthermore, the bulk of CPS engages sophisticated response controller expertise. Response management denotes the capability to control cyber-physical measures in response to modifications in the physical environment [7]. The CPS is an important part of the Industrial Internet of Things (IIoT). Smart programs and services can function in real time and with pinpoint precision thanks to CPS. They are constructed on the actual exchange of variety of data and essential information through the integration of cyber and physical systems [8].

Adopting industry 4.0 even without an external entity, which alters by combining new technologies and systems, delivering independence, managed networked, state-driven art accessibility, and reliability [9].

Biometrics, big data, smart grid, Internet of Things (IoT), neural networks, and deep learning are only a few of the major topics covered by CPS. IoT devices have recently become a serious concern, boosting the platform's vulnerability to cyber assaults. To handle cyber and physical risks, such platforms must be built with a secure, reliable, and fault-tolerant framework [10].

CPSs have a variety of weak points where cyber physical attack vulnerability compromises, resulting in data breaches to control platform dependability and robustness [11].

In today's environment, CPS security needs a lot of effort since a growing number of devices are affected during information transit [12].

The safety and security requirements of CPS devices need careful consideration in both architecture and research [13]. Physical process, networking, computing, and actuation are the four phases of the CPS communication process. The CPS workflow is depicted in Figure 9.2.

(i) Physical process: The primary purpose of this phase is to monitor the CPS and to build a connection and process communication channel. It offers data on prior values and ensures feedback on previous actions as well as proper operation in the actual world. The sole purpose of physical processes is to ensure resource integrity and appropriate operation.

(ii) Networking: Data aggregation and dissemination must be possible throughout the networking phase. The CPS sensor used to be able to store a large quantity of data. Such sensors provide records in real time and have a large number of sensors that can take data for analysis to evaluate forthcoming necessities.

(iii) Computing: Information gathered during the present stage is mostly utilized for interpreting and evaluating the level of complication being supervised, as well as ensuring that the physical process meets pre-determined standards. If

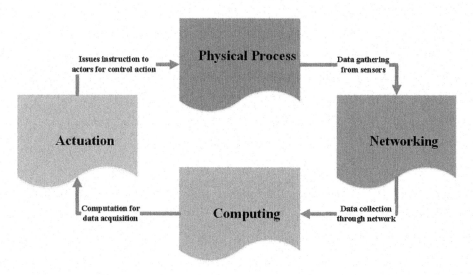

FIGURE 9.2
CPS workflow phases.

satisfactory results are not obtained, remedial action is suggested to attain the desired outcomes.

(iv) Actuation: The findings of the computation phase are used in this step to activate specific functionality.

9.2 Architecture of Cyber Physical System

Computing systems, internetworking, and physical processes are all integrated into cyber physical systems. As shown in Figure 9.3, these designs are usually characterized by three primary levels: physical, network, and application layers.

(i) Physical layer: Physical layer consists of sensors/actuators and a range of supplementary devices and subcomponents having sensing/processing and communication abilities [14]. Sensors inside this layer capture physical processes in the platform's physical surroundings. Actuator elements, that react to instantaneous event monitoring and interface with the application layer to facilitate data processing, are also included [15, 16]. The actuator units have the capacity to alter the characteristics of real-world physical objects and events [17]. Cyber physical activities are created as

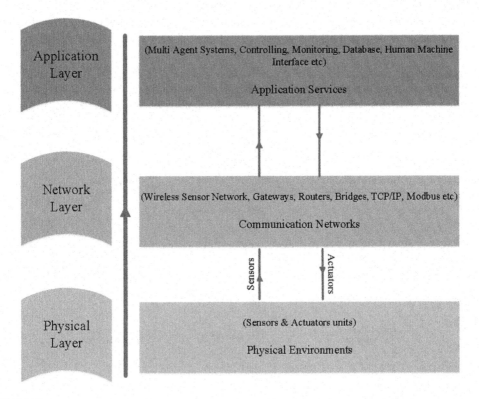

FIGURE 9.3
Architecture of a cyber physical system.

a result of the platform's processing of information, and they induce a physical process to be triggered. A component of this layer has the capacity to communicate with other networks, such as the internet, using a gateway is a key feature. Implementing security guidelines is critical since this level is so prone to cyberattacks.

(ii) Network layer: The network layer's job is communicating control instructions and sensual data b/w the applications and physical layers. Because there are so many distinct diverse networks connecting, specialized security procedures must be considered. Although it acts as a connection between the physical and application layers, as well as a pathway between sensors/actuators, network layer is important to CPS functioning [18, 19]. The network layer is the path via that information, sensor evaluation and orders to actuation are sent and received. Ethernet, Distributed Network Protocol (DNP), Preferred Specification 16(RS-232), TCP/IP, dial-up modem, and various wireless protocols are examples of network layer protocols [20]. Wireless communication is the most preferred technique for sensing and actuation communication due to its dispersed landscape. Aimed at wireless technology, cloud or actual servers are considering commonly. Network layer links application and physical levels, capable of communicating and collaborating effectively.

(iii) Physical layer: Several features make up the application layer, including control systems, databases, and an interface design. This layer collects input from the network layer and creates control orders for actual machines and systems to be controlled [21]. Access control, precision farming, remote sensing, smart cities, electric grid, and other application-specific roles include data accounting and depiction, trying to act as a controller for end customers, and proposing most visual interface for a wide collection of applications/services are all handled by the application layer. It also gives end users an interface and services for obtaining sensual data through portable items or workstations that might take many dissimilar procedures and consume very different safety needs.

9.3 Security Issues and Challenges of Cyber Physical System

CPSs are at a pivotal point in their growth, and as a consequence, they confront a slew of challenges, the most significant of which is security. CPSs are subject to cyber assaults aiming at gaining interior information otherwise interrupting storage and processing, much as conventional software systems [22]. Intruders try to obtain access to the system, distribute harmful code or malware, or collect confidential material for their malevolent purposes, like harming businesses or impersonating a genuine user by collecting personal information [23]. When assaults against CPSs cyber aspect are carried out, the ability which regulates cyber physical events are disturbed [24]. Because they execute many actions at various stages, CPS must protect appliances, information transfers, applications, memories, and actuation procedures. The requirements are:

(i) Securing device access: The first difficulty is protecting device access. If authorization is not given or is supplied in a bad manner, unauthorized entities may gain entry and alter the environment [25]. As a consequence, neither the integrity of the underlying binary codes nor the software functionality can be assured.

(ii) Securing transmissions of data: Data transmission security is critical for detecting pretenders and malicious acts in CPS communications systems, as well as restricting unauthorized access. For example, studying the data sent and received, attackers try to capture the physical characteristics of network energy usage and temporal characteristics [25]. Certain intruders seek and interrupt systems by initiating DoS attacks or altering route architecture [26]. Several terminal devices lack significant processing of data, communication, and data storage since they're not complete systems [27]. As a result, cyberattacks are much more likely to affect these devices. Interconnection in Industrial Control System interfaces improves operational efficiency and decreases operating costs by relying on open communications protocols. While these interfaces make operations more efficient and productive, they also expose the infrastructure to additional intrusions and hostile assaults, such as malicious software, DDoS, eavesdropping, and unapproved access [28]. Another contributor to security vulnerabilities is described as even design procedures are constantly forced in terms of computational acceleration, physical servers, and energy usage. Moreover, integrated structures are typically intended by specialists with little expertise with security risks, and often prioritize functionality, checking errors, and security throughput [29]. As a result, the platform is susceptible and confidential material might be exposed to unauthorized or undesired individuals.

(iii) Application security: The application layer is where a field of application, and also security issues, are brought all together. The data privacy that occurs at this layer would not be resolved at another level, which does not face the same security challenges. As a result, hackers have access to critical information about users, which can lead to information breaches and privacy issues. Spatial camouflage, faceless space, and spatial cryptography are certain data security techniques employed at this layer since this data may reveal information about prior and present places visited by users. Many apps on this tier also have an influence on customers' social life, demanding their security [30].

(iv) Data storage security: It's critical that private data in CPS equipment is kept secure. Sensors, for example, make up the bulk of CPS equipment [26]. They are tiny, wireless-linked nodes with limited resources. Regardless of the fact that several technology approaches employ cryptography methods for certain instruments, storage restrictions and the devices' restricted computing power make them inadequate. Consequently, light-weight security techniques are essential [24].

(v) Securing actuation: Any actuation actions must originate from a trusted source, according to actuation protection. In the event of a threat, this will guarantee that the response and management orders are exact and protected. [27]. Accordingly, to use the web as a transmission layer within CPS interconnections, there'll be security issues. Security must be given for entire system equally unified endwise defense policy [30], instead of individual functional security tool at every level. Furthermore, any desired security system must meet large memory requirements and perform heavyweight calculations [28].

9.3.1 CPS Security Threats

According to the architecture of the cyber physical system (CPS) shown in Figure 9.3, security threats to the CPS may be distributed into three categories, physical-layer threats, network-layer threats, and application-layer threats:

(i) Threats in physical layer: Within the multilayered architecture of CPSs, the physical layer will be the significant foundation of sensory data and implementation location for control orders. Because the bulk of physical layer nodes in the network is located in unsupervised contexts, they are convenient targets for hackers. So the physical layer's observation network has limited data analysis, transmission, and storage capability, traditional security approaches cannot be directly implemented. The physical layer's main security threats are detailed in Table 9.1.

The physical security of every node's architecture, the gathering of represented data, and the implementation of control instructions are all considerations at the physical layer. Sensors/Actuators, RFID instruments, picture capture devices, as well as additional technologies should be safe thanks to this layer's security countermeasures. The physical layer's security is the bedrock of cyber physical system security. The physical layer's major security mechanisms and countermeasures are as follows:

- Improve the administration and security of the node's identification. This will, to some extent, extend the time it takes for a node to be approved. In real implementations, admins can evaluate the system's effectiveness and safety to develop a more appropriate node authentication approach.

TABLE 9.1

Security Threats at Physical Layer

Security Threats	Description
Physical assault	Physical attacks mostly called to the concrete damage of the nodes.
Failure of equipment	Equipment functionality is reduced or lost as a result of external pressures, the surroundings, or age.
Electric line failure	Line failure refers to power lines failure on nodes.
Electrostatic leakage	Intruders can recover actual information by analyzing electrostatic communication devices emitted out at work.
Electrostatic intervention	Excessive electromagnetic communications or commotions degrade overall performance by interfering with valuable signals.
Denial of service (DoS)	The intruder leads the target computer to stop providing services by consuming network bandwidth.
Channel blocking	Data transmission is not possible due to a transmission medium that has been saturated for a long period.
Sybil attack	A unique rogue node utilizes several identities to assault the network by commanding the majority of nodes.
Data spoofing	An intruder intercepts and changes data before sending it to the intended receiver.
Unauthorized access	Malicious hackers get access to resources.
Passive attack	Sniffing and information gathering are used by the attacker to passively acquire data.
Entity capture	An intruder gets hold of a gateway node or a normal node.
Replay attack	The hacker retransmits the genuine data received before in order to gain the system's confidence.
Perception data destruction	The unlawful insertion, deletion, alteration, or destruction of perceptual data.
Data intercept	Obtaining unauthorized access to information resources by intercepting a line of communication.

- Biometric systems and relatively close to digital systems will be used to better protect perceptual data.
- Make the legislation more robust. Make the consequences of unlawful activity with cyber physical systems obvious.
- Cyber physical systems combined with cryptography [17, 18], privacy protection [19–21, 26, 31–33], security routing [34, 35], security data fusion [22, 23, 36], and safety positioning [24, 25] research.

(ii) Threats in network layer: The "next-generation networks" acts as the main carrier network for CPS network layer. The design, access techniques, network infrastructure of the "forthcoming generation network" will expose CPS to the multitude of security vulnerabilities. The high number of nodes and enormous information over the network layer might introduce congestion issues, rendering the structure vulnerable to DoS/DDoS attempts. As a result of information exchange, entry point authorization, and the convergence of security rules across various networks, additional security problems for the network layer of CPSs will emerge. The major network layer security threats are shown in Table 9.2.

The network layer of CPS has security mechanisms in place to certify integrity, confidentiality, and uniformity of transmission information within the systems. For network layer protection, both peer-to-peer and endwise cryptography methods can be employed [26]. The network layer's major security mechanisms and countermeasures are as follows:

- A point-to-point cryptography method ensures data confidentiality throughout hop-by-hop communication. This approach demands a greater degree of node dependability since any node in the chain may acquire plaintext data. A few of the safety features include peer authorization, node-to-node encrypting, and inter-network verification.
- Endwise encryption provides data protection from start to finish, with adjustable protection rules for different levels of security. Edge encryption, on the other hand, may not hide the data's sender/receiver, and enemies might utilize that knowledge to their advantage. Some of the security techniques include endwise verification, important negotiating, and access control.

(iii) Application-layer threats: The network layer of CPS has security mechanisms in place to confirm the integrity, confidentiality, and uniformity for transmission of data in the system. For network layer protection, both steeplechase and endwise cryptography methods can be employed [26]. This approach demands a greater degree of node dependability since any node in the chain can acquire plaintext data. A few of the safety features include peer authorization, hop-by-hop encrypting, and inter-network verification, the description of security threats are shown in Table 9.3.

The application layer is the most important part of a CPS, since it is responsible for creating choices and giving control orders. Huge data amounts on the application layer of CPSs demand sophisticated data processing skills and comprehensive user privacy data security on this layer. The following are the primary physical security measures:

- Strengthen the platform's access control list.
- In various circumstances, strengthen the authentication and encryption mechanisms.

TABLE 9.2

Network Layer Security Threats

Security Threats	Description
DDoS	A huge number of malicious nodes assault the target server around the same moment as DoS sources.
Routing attack	The intruder sends counterfeit routing information to disrupt the regular routing procedure.
Sink node attack	By targeting the sink node, data transfer between the physical and network layers is disrupted.
Direction misleading attack	Malicious node changes senders and receiver's addresses of packets of data before sending them down the incorrect path, causing confusion in network routing.
Black hole attack	A malicious node deceives neighboring nodes into establishing routing links with it, then discards packets that should have been routed, resulting in packet loss.
Flooding attack	Smurf and DDoS consume the network servers' capabilities upon that network layer.
Trapdoor	Permit an exemption to the security policy for transmitting particular data.
Sinkhole attack	A malicious user draws regular nodes to itself as a point in the transmission range, causing data to communicate via it.
Wormhole attack	Malicious nodes band together to improve routing by reducing the number of routing hops among them.
Routing loop attack	A malicious node alters data route, resulting in endless routing loops.
Sybil attack	By owning the most of the terminals, the rogue node can impede data transmission by using several identities.
HELLO flooding attack	The malicious node notifies others in the network which are their immediate neighboring via broadcasting routing information through a strong signal.
Spoofing attack	A malicious node impersonates a legitimate node in order to send data along a sluggish route or to a node that has failed.
Selective transporting	A malicious node loses part or every critical information in the redirect process on purpose.
Tunnel attack	Malicious nodes conceal their true connection extent in order to entice additional nodes to create routes via them.
Fake routing details	A malicious node tampers with the route knowledge to exploit the network layer system.

TABLE 9.3

Application-Layer Security Threats

Security Threats	Description
Privacy data leakage	Consumers' personal information is exposed due to unsecured data transfer, storage, and display.
Illegal access	Connectivity to network and system data that is not permitted.
Malicious code	The system contains code that has no impact but may be a security risk.
Stolen control instructions	Intruders deliberately utilize or damage a system by changing control commands.
Loophole	Attacking the system through application-layer flaws.
Viruses and Trojan horses	The most frequent application-layer security risks are viruses and Trojan horses.
SQL injection Attack	SQL injection is a typical method of hacking a system's database.

- Strengthen network forensics capabilities by improving network forensics mechanisms [27].
- Create consolidated, well-organized security management for different cyber physical system applications.

9.3.2 CPS Security Attacks

Attacks against CPS might result in substantial environmental harm. Passively or aggressively, each layer of the CPS can be assaulted. CPS is also more prone to assaults than traditional IT systems. These assaults aren't limited to CPS; they may also target the network as a whole, notably the internet [23], in use as the transmission layer. Physical layer intrusions involve assaults on nodes like sensors and actuators; network layer attacker uses data leak or destruction, as well as security flaws throughout data transfer; intrusions involve illegal entry that violates user privacy [37] at the application level. As a result, identifying possible risks and putting in place solid security architecture is critical. Regardless of the fact that all layer is prone to adverse sorts of assaults, certain assaults may hit all levels, and examples of these attempts according to [23, 26, 37, 38] include:

- Denial of service (DoS): It modifies performance using limiting communication in terms of making the networking and service inaccessible, for example, by overwhelming assets with fake demands and exploiting a protocol flaw. DDoS is also a public assault that victimizes many resources at the same time, such as end devices and networks, preventing accessibility and services [38].
- Man-in-the-middle (MITM): It delivers a false message to a victim's assets that subsequently takes excluded activities in response to the communication, such as manipulating an original purpose, which might result in an unpleasant occurrence. This sort of attack can also damage the network layer, which in some cases can lead to eavesdropping [39].
- Eavesdropping: The system's data is captured at all times. Transferring control data from sensor networks to CPS applications for monitoring reasons, for example, might be subject to eavesdropping. Additionally, user privacy may be jeopardized so the system is being watched [39].
- Spoofing: When trying to participate in system activities, pretends to be a genuine customer. The intruder will have access to information and will be capable of changing, removing, or inserting data after successfully acquiring access [39].
- Replay (playback): To gain the system's confidence, retransmits data received from a destination node [38]. This sort of attack is carried out by spoofing and changing or reacting to one of the devices' identifying information.
- Compromised key: This assault is aimed at the secret key which can be used to secure messages. A timing (side channel) exploit is used to do this, that includes calculating the needed encryption time [38]. The hijacked key would then be used to alter data and conduct computer investigation to decode additional private keys in similar system. In rare circumstances, attacker might acquire access to sensors to complete business operations in order to extract additional internal keys. During different situations, an intruder might be able

to replace a sensor node and pass themselves off as the genuine edition while swapping keys with other units [26, 27], revealing the additional nodes' underground keys.

At each level within CPS, various sorts of hazards and common assaults for each layer may be categorized as follows, based on the CPS design:

(i) Attacks at the physical layer: End devices such as RFID tags and sensors make up the physical layer, which is constrained by computer resources and memory capacity. Furthermore, because such items are typically used in exterior and outdoor environments, they are susceptible to physical assaults such as component tampering or replacement. As a result, terminal devices are particularly vulnerable to numerous types of assaults. Physical layer attacks include device malfunctioning, link failure, electromagnetic conflict, cognitive information skimming [23], discrepancy power investigation [38], information disclosure, monitoring, forgery, detecting security breaches, complete destruction, and power fatigue. Examples of frequent sorts of assaults include the following:

- Node capture: Gets hold of the cluster and acquires and exposes data, including encryption keys, and formerly utilized jeopardize whole system security. Secrecy, availability, veracity, and legitimacy are all targets of this type of assault [38, 40].

- False node: It introduces a new node to the network and broadcasts malicious data to jeopardize the integrity of records. It can induce DoS attack via depleting power of platform's node [38].

- Node outage: Halts node operations, making it tough to declaim and collect data from them, and initiates a slew of other assaults that jeopardize the network's availability and integrity [40].

- Path-based DoS: Overflowing packets are dispatched with a transmitting path to the origin station, depleting node cells and disturbing the network, decreasing node uptime [40].

- Resonance: Forces sensors or controllers that have been tampered with to operate at an alternative resonance frequency [41].

- Integrity: Efforts to disrupt the system by manipulating sensor data and introducing external influence inputs [42].

(ii) Attacks at the network layer: Data leakage during data transfer is one type of attack against another layer. This occurs as a result of the communication medium's accessibility that is particularly true with wireless systems. These exploits imitate digital credentials by collecting a broadcast message through a radio channel, changing and demodulating it, or moving information across composite networks. Additional characteristics, including such web access methods used by a significant set of network endpoints, which could also generate congestion problems, could render it more vulnerable to assault [23, 39]. Reaction and manipulation, exploitation, fatigue, crash, black hole, flooding trap doors, base station, orientation confusing sinkhole, wormhole, improper route planning, tunneling [23, 26], and illegal access are all prominent assaults at this layer. Here are a few examples of common network layer attacks:

- Routing: Routing loops can cause network transmission resistance, higher transmitting latency, or a longer source path [26, 38].

- Wormhole: It creates information gaps in the network by constructing fake pathways through which every container is transmitted [43]. An announcement of routing to supplementary nodes is made. Other attacks, such as selective forwarding and spoofing, might be launched using this method [26]. Introducing noise or a signal of comparable frequencies into the communication networks between the network system and the faraway ground control station effectively jams the connection. By creating deliberate network disturbance, this attack might result in DoS.

- Selective forwarding: Allows hacked endpoints to discard packets while forwarding those that are chosen. While the hacked node is still considered authentic, it may cease sending packets to its intended destination or selectively relay certain messages while rejecting the rest [26].

- Sinkhole: For traffic routing to other nodes, the optimal routing path is announced. This technique could be used to initiate further attacks, like selective forwarding and spoofing.

(iii) Attacks at the application layer: Because it collects a considerable amount of user data, attacks upon the layer lead to data loss, confidentiality loss (such as user habits and health conditions), and unauthorized equipment. Customer data loss, illegal access, malware, record forging, and control command fraud are all examples of application-layer risks [23, 24]. Assaults on this layer are commonly seen as follows:

  - Buffer overflow: Attacks are done by utilizing the advantage of every vulnerability that may lead to a buffer overflow [38].

  - Harmful code: Attacks the customer's application by generating harmful script, like viruses and worms, which slows or damages the network.

9.4 Security Objectives of Cyber Physical System

When it came to guaranteeing the protection of CPSs, there are a few security goals to achieve:

(i) Confidentiality: Confidentiality refers to a person's or system's capacity to keep information from being shared with unauthorized persons or systems [44]. Take, for example, a Healthcare CPS. By encrypting the individual health history throughout communication, restricting the locations where it would surface (in datasets, system logs, archives, etc.), and limiting access to the places in which it is maintained, the system intended to guarantee secrecy. When an illegal entity has access to personal health information, this seems to be a violation of confidentiality in some form.

(ii) Integrity: In order to preserve integrity, CPSs must be capable of accomplishing physical goals [27] by preserving, detecting, or suppressing manipulation risks on information- supplied and acquired by sensors or controllers [45]. The capacity to detect any (malicious purposes) introduced changes in the message being transmitted is needed to ensure data integrity [46].

(iii) Availability: A CPS's high availability attempts to guarantee that operations are often reachable by avoiding computation, management, and transmission corruptions induced by component failure, efficiency improvements, power cuts, or DoS [47] presented a multi-cyber infrastructure based on the Markov model to increase CPS availability.

(iv) Authenticity: In computer and transmission operations, it requires confirming the authenticity of data, communications, and conversations. Authentication of all linked processes, such as sensing, communications, and actuations is the objective of CPS.

(v) Trustworthiness: This is the degree to which users may trust the CPS to complete needed activities within specific domain and time restrictions. The software, hardware, and collected data must all exhibit a level of trustworthiness in order for a CPS to be viable and trustworthy.

9.5 Proposed Methods and Approaches to Secure Cyber Physical Systems

CPS assaults are more difficult to detect and prevent than internet attacks. Hackers may use numerous attack phases to get access to a CPS in order to avoid detection. In the sections below, a number of security techniques for dealing with CPS assaults are detailed. The CPS security techniques and their relevant references are shown in Figure 9.4.

(i) Attack taxonomy:

Cyber assaults against CPSs include not just typical IT-related cyberattacks, but also attacks on CPSs that can transcend the cyber physical domain border [48]. The taxonomy of assaults against CPS has already been addressed in depth in the above section.

(ii) Attack and threat modeling:

FIGURE 9.4
Modeling security approaches: cyber physical systems.

Due to a number of constraints, simulation and modeling are commonly used to do research on CPS attacks. The importance of replicating real-world assaults cannot be overstated [49] [50]. Proposed hybrid attack graphs as a novel approach to modeling cyber-physical risks in smart grids. Fabio Pasqualetti et al. proposed an attack paradigm that integrates stealth and replays assaults into a unified architecture with better monitoring systems for malfunction and attack detection [51]. Michael E. Kuhl and colleagues proposed a cost-effective and time-saving modeling technique for representing computer networks in order to simulate cyberattack scenarios for the purpose of assessing security solutions [52].

Because heterogeneity, dynamism, and complexity demands must be properly investigated and addressed, CPS modeling is important [53]. From information transmission to process modeling, CPS modeling may cover a wide range of issues [54]. Security techniques such as threat modeling, strategy planning, and a holistic approach to tackling security and safety concerns together are discussed in this context by [55]. There are several problems that apply to all models, regardless of their nature. The model should be deterministic, capable of aiding in problem-solving, provable, and executable [56]. System dynamics, Bayesian networks, coupled component models, agent-based models, and knowledge-based models are five typical modeling techniques for a complex system, according to [57]. A Bayesian network's objective is to characterize the influence of one entity's attributes on the characteristics of another entity, or one event on another event. The linked component modeling method, on the other hand, entails integrating models or model components from a variety of fields to arrive at a comprehensive solution [57, 58]. All system elements are represented by interacting agents with particular behavior that takes into consideration their influence on the overall system in agent-based modeling. Finally, utilizing a knowledge base and logic tools, knowledge-based modeling extracts solutions.

(iii) Attack detection:

Several researchers have looked at the detection, isolation, and recovery of control system faults. [59]. CPSs have vulnerabilities that are not present in traditional control systems, and suitable detection and identification methods must be developed for them [60]. Physical dynamics-focused internal assaults fail because security measures don't conform to the underlying physical process or method of control in any quantifiable way [61]. Robert Mitchell et al. developed a general hierarchical model for assessing the performance of intrusion detection algorithms in the context of a CPS [62]. They create two intrusion detection techniques for identifying fraudulent attacks in a CPS and then use the hierarchical model that was created to assess the results of the two techniques and determine the best design parameters for improving the CPS's dependability. The author suggested and tested a behavior regulation, configuration-based exploitation detection method for medicinal devices surrounded in healthcare CPSs [63]. The causal connection between components in a CPS is modeled using Bayesian networks and casual event graphs [64]. Florian Dorfler created a unified prediction system and an enhanced detection approach in order to prototype an assault on a core network [65]. An Intrusion Detection Device (IDS) is a vital component of cyber surveillance equipment that is designed to alert either a human operator or another information security equipped system when an assault occurs. The primary functions of IDS in CPS are to gather and analyze data about the entity that has been infiltrated [66, 67].

- Network-based Protocols: Focuses on network traffic and the dangers it poses, taking network protocols, traffic, and devices into consideration.

- Wireless: Identical to the first, except that those IDSs include wireless traffic and protocols.
- Network behavior analysis: Monitors network traffic flows to identify unusual activity patterns and policy breaches.
- Host-based IDSs monitor traffic, application activity, file actions, and configuration activities for a single host. This type of ID is commonly used on critical infrastructure nodes.

Based on how they identify threats, modern IDS may be divided into three categories [68]:

- Anomaly-based approaches presuppose the identification of deviations in behavior from the normal.
- Signature-based systems depend on a database of up-to-date threat models to identify and prevent intrusions.
- The identification of questionable behavior across the whole system, down to the level of individual subsystems and interconnections, is a fundamental tenet of specification-based approaches.

In [68], the authors provide a unique approach of classifying IDSs by classifying them according to the audit material they examine, such as the host or network, and the detection methodology they use, such as experience and understanding or behavior patterns. The same research also clarifies the distinctions between standard IDSs and IDSs for cyber physical systems. To deal with the problem of CPS resource limits, scheduling techniques have been devised, in which responsibilities are shared among CPS nodes in order to keep the processing flow constant [69] [68, 70]. It offers a plan to improve the likelihood of an attack being recognized by the use of verified safe monitors. Data obtained from a reference device is compared to data collected from other devices in the same environment using so-called trustworthy devices [71]. Using this concept assumes that a collection of sensors measure the same underlying physical parameters. The measurements are then collected and evaluated using a sensor algorithm that takes into consideration the sensor precisions that have been established. However, this technique is not particularly practical and may expose the system to both passive and DoS assaults. Time-based detection is another method of intrusion detection in CPS [72], which assumes a predefined variable for worst-case execution time (WCET). If the execution time exceeds the preset limits, the system is considered hacked.

(iv) Security solutions

Keeping protected CPS surroundings is not stress free due to the consistent growth of issues, integration concerns, limitations of presenter solutions, such as a security/privacy/accuracy deficiency. Cryptographic and non-cryptographic techniques are included.

(v) Cryptographic-based methodologies

These protections are common in SCADA systems to secure the route from active and passive attacks, and also illegal access and disruption [73]. The main goal must be on protecting and ensuring the efficiency of the entire production system, rather than on data security. As a result, a range of choices were made available [74]. Investigated CPS security and the degree to what reference images may be designed to improve safety. Security problems, flaws in present security architecture, and alternative ways to restoring security in power grids are all examined [75]. Security risks to water distribution systems are highlighted [76], along with the need to build threat prototypes and supervision roles in security management.

(vi) Non-cryptographic-based solutions

There were also a number of non-cryptographic solutions offered to limit and eliminate any possible cyberattacks, including the following:

1. Intrusion detection systems (IDSs): Different IDS approach types are accessible due to the availability of diverse network topologies [77]. When it comes to detection, setup, pricing, and network location, each IDS approach has its own set of benefits and cons. A number of research projects were undertaken in order to identify CPS assaults. [78]. These assaults are divided into two categories. A physics-based paradigm for defining basic CPS operations using anomaly detection in CPS. According to [79, 80], a cyber-based model is utilized to detect possible assaults. In reality, the techniques described were designed primarily to identify particular attacks against specific applications, such as industrial control processes [81] and smart grid [75].

2. Firewalls: Since the introduction of intrusion detection systems (IDS) and artificial intelligence (AI), firewalls have seen minimal use in the CPS sector. As a result, only a few firewall-based solutions are available [82]. With the goal of improving server computer networks, this report describes the use of paired firewall in various network zones. The authors picked paired firewalls because of the rigorous security and clear administrative separation. Ghosh et al. described how they used event data to predict real-time network device failures such as firewalls [83].

3. Honeypots and methods of deception: CPS uses deception as a decoy to conceal and defend its system, which is a crucial defensive security measure. Honeypots are mostly used to do this. Cohen talked about how honeypot deception might be enhanced when used. [84], employing a variety of deceptive techniques. Author proposed in [85] the HoneyBot was created using the HoneyPhy architecture. According to the research, HoneyPhy may be used to replicate these actions in real time. According to a recent study, HoneyBot can fool attackers into thinking their attacks are effective.

4. Machine learning, threshold, and rule-based schemes: Understanding the distinction between basic rule-based procedures and machine learning algorithms is critical (ML). At the same time, both can be represented in the same security architecture. Rule-based mechanisms, which are based on previously collected information, are used as the first line of defense against threats, detecting the most significant risks that might harm the system. Feature extraction, feature selection, rule-based detection, and ML algorithms for threat detection and rule extraction are all important elements in a security architecture that combines ML algorithms with rule-based processes. A more sophisticated method is to incorporate human supervision into the security architecture, as provided by extracted information from the machine-learning module. Specific traffic types, such as P2P, may be

prioritized from the raw input, which could be represented by packets or log files that are recorded and analyzed. The identification of critical characteristics that are useful for threat detection is part of the second step. In the third step, rule-based processes are applied to the list of important traits, and potentially hazardous elements are discarded.

5. Knowledge-based schemes: The ML module checks other entities that have not been identified as harmful after the rule-based detection module has removed malicious entities and created the learning set for the ML module. In addition, rule inference modules can be added to the ML module to produce new rules in response to new threats.

6. Security architecture and design:

Networked resource-contained devices are expected to make up the bulk of CPSs. CPS security is essential for preserving data integrity and privacy while bolstering system defenses against intruders. Sebastian Seeber and colleagues illustrate how security may be utilized to secure communication in an RPL network in [86]. New security risks would target CPSs, such as the ability to manipulate the structure at the IT system level and inside its immediate environment. In [87], the author delves into the many security flaws that exist, and suggests a novel system design that merges ideas from the realm of Biological Computing with the goal of developing upcoming safe CPS.

9.6 Security Research Areas in Cyber Physical System

In the context of CPS security, this section contributes to a discussion of many major outstanding research problems.

- Protection of critical infrastructure: Smart grids are an excellent example of essential infrastructure, with several significant issues such as
  - (i) Variety of technologies and protocols;
  - (ii) Communication protocol flaws, since some lack security measures;
  - (iii) According to [73], certain physical devices have restricted capabilities;
  - (iv) A wide range of security approaches that vary depending on the application domain.
- Cyber and physical components protection: One of CPS's most essential criterion is integrity. The security of a sensor network, as well as its superstructures and data integrity, must all be taken into account. Another important challenge is the lack of a common technique for developing secure CPS, which has resulted in a plethora of proprietary solutions based on possibly flawed methodologies.
- Security and mobility: Because mobile devices connect with so many other networks and services, they are potential carriers of threats. Increased security protections for wearable smart mobile assistants and implants have become a serious issue since these gadgets have the potential to threaten human health and even life.

- Safety communication for geographically dispersed points: Ensuring secure communication for certain of a system's critical nodes or subsystems is a challenging undertaking. Furthermore, apps that access data collected by sensor nodes, for example, might be hijacked, resulting in a hard-to-detect passive attack on the sensor network. The notion of federated learning, in which data is processed locally rather than routed to a central node for processing, might be adopted as a viable strategy. As a result, the local data set does not need to be sent to a central server; just a global model update is required [88].

- Analytical and integration tools: There exists an uncertain necessity for novel security models which integrate machine learning, decision-making algorithms, and human-assisted analysis. Human-assisted analysis is still an important part of recognizing dangers that have already been experienced. In addition, because the most common threats demand real-time countermeasures and data volumes are enormous, new solutions for quicker response times are necessary.

- Proactive security systems: Based on prior threats, proactive security systems enforced by analytical tools might identify major attack paths and threat types. A DDoS assault, for example, may be the most prevalent hazard to a system, necessitating special protections.

- Enabling collaborative mechanisms for CPS protection: A shared threat mitigation plan, as well as information exchange about previously experienced dangers, are examples of collaborative techniques. A fragmented security approach is no longer feasible against emerging threats, according to a McAfee research [89]. A proposal for an open integrated ecosystem for security activity coordination was also developed. As a result, the sharing of threat information will assist security systems belonging to a variety of stakeholders. As new threats develop quicker than threat mitigation systems can respond, reaction time and resource requirements will be reduced. Another approach is to employ a common knowledge base that is being expanded by several parties in order to increase threat mitigation efficiency.

9.7 Conclusion

Complex methods that rely on the convergence of physical and cyber or software components are known as CPSs. CPS provides a diverse set of services. The number of CPS deployed is steadily rising, posing a multitude of security and safety concerns. In this study, researchers take an in-depth look at the idea and architecture of cyber physical systems (CPSs), discuss the security goals and challenges associated with CPSs, and analyze the vulnerabilities and attacks faced by three CPS deployment tiers. It also spoke about how important it is for CPS to develop new sophisticated methods in the field of security and privacy. We looked at several mitigating strategies, such as sophisticated intrusion detection systems that use machine learning algorithms. The fast growth of CPS has had a profound impact on our way of life. CPS is now the highest focus in order to improve its capabilities and create a lifecycle as easily as possible. Such CPS equipment is strongly coupled by cyber space components and physical devices and are unified and synchronized with the physical surroundings. CPSs are a viable framework for creating current and prospective systems, and they have been predicted to get a significant influence on the actual world. CPS is more concerned with the design of complicated processes as a whole, instead of particular cyber or physical systems. CPS systems are essential components of Industry 4.0, enabling Industry v4.0 by integrating the physical and cyber worlds. They're clearly changing the way individuals interact with their environment. CPS systems have a number of security and privacy problems that might jeopardize their dependability, safety, and efficiency, as well as prevent broad implementation. In order to give a theoretical reference, we looked at a variety of CPS risks and assaults in this chapter. We also looked at the cyber physical system's security objectives, ways to secure CPS, and security problems, as well as open research questions. The CPS environment requires safe infrastructure to resist severe cyber-physical attacks. We anticipate that CPS security researchers will find this article useful in a variety of ways.

References

1. Gurgen, L., Gunalp, O., Benazzouz, Y., & Gallissot, M. (2013). Self-aware cyber-physical systems and applications in smart buildings and cities. In *Design, Automation and Test in Europe Conference and Exhibition (DATE)* (pp. 1149–1154). IEEE Publications.

2. Rajkumar, R., Lee, I., Sha, L., & Stankovic, J. (2010). 47th ACM/IEEE. Cyber-physical systems: The next computing revolution. In *Design Automation Conference (DAC)*. (pp. 731–736). IEEE Publications.

3. Da Xu, L. D., He, W., & Li, S. (2014). Internet of things in industries: A survey. *IEEE Transactions on Industrial Informatics*, 10(4), 2233–2243. https://doi.org/10.1109/TII.2014.2300753.

4. Gubbi, J., Buyya, R., Marusic, S. P. M., & Palaniswami, M. (2013). Internet of Things (IoT): A vision, architectural elements, and future directions. *Future Generation Computer Systems*, 29(7), 1645–1660. https://doi.org/10.1016/j.future.2013.01.010.

5. Wan, J., Yan, H., Suo, H., & Li, F. (2011). Advances in cyber-physical systems research. *KSII Transactions on Internet and Information Systems*, 11, 5.

6. Brak, M. E., Brak, S. E., Essaaidi, M., & Benhaddou, D. (2014). Wireless sensor network applications in smart grid. In *International Renewable and Sustainable Energy Conference (IRSEC)* (pp. 587–592). IEEE Publications.

7. Bordel, B., Alcarria, R., Robles, T., & Martín, D. (2017). Cyber–physical systems: Extending pervasive sensing from control theory to the Internet of Things. *Pervasive and Mobile Computing*, 40, 156–184. https://doi.org/10.1016/j.pmcj.2017.06.011.

8. Lee, J., Bagheri, B., & Kao, H.-A. (2015). A cyber-physical systems architecture for industry 4.0-based manufacturing systems. *Manufacturing Letters*, 3, 18–23. https://doi.org/10.1016/j.mfglet.2014.12.001.

9. Zeng, J., Yang, L. T., Lin, M., Ning, H., & Ma, J. (2020). A survey: Cyber-physical social systems and their system-level design methodology. *Future Generation Computer Systems*, 105, 1028–1042. https://doi.org/10.1016/j.future.2016.06.034.

10. Liu, C., & Zhang, Y. (2016). *Cyber-Physical Systems*. CRC Press.

11. Akella, R., Tang, H., & McMillin, B. M. (2010). Analysis of information flow security in cyber–physical systems. *International Journal of Critical Infrastructure Protection*, 3(3–4), 157–173. https://doi.org/10.1016/j.ijcip.2010.09.001.

12. Wang, L., Törngren, M., & Onori, M. (2015). Current status and advancement of cyber-physical systems in manufacturing. *Journal of Manufacturing Systems*, 37, 517–527. https://doi.org/10.1016/j.jmsy.2015.04.008.

13. Sobhrajan, P., & Nikam, S. Y. (2014). Comparative study of abstraction in cyber physical system. *IJCSIT International Journal of Computer Science and Information Technologies*, 51(0975–9646), 466–469. Accessed March 7 2020.

14. Estrin, D., Culler, D., Pister, K., & Sukhatme, G. (2002). Connecting the physical world with pervasive networks. *IEEE Pervasive Computing*, 1(1), 59–69. https://doi.org/10.1109/MPRV.2002.993145.
15. Oliveira, L. M., & Rodrigues, J. J. (2011). Wireless sensor networks: A survey on environmental monitoring. *JCM*, 6(2), 143–151. https://doi.org/10.4304/jcm.6.2.143-151.
16. Wang, S., Wan, J., Li, D., & Zhang, C. (2016). Implementing smart factory of industry 4.0: An outlook. *International Journal of Distributed Sensor Networks*, 12(1), 3159805. https://doi.org/10.1155/2016/3159805.
17. Dargie, W., & Zimmerling, M. (2007). Wireless sensor networks in the context of developing countries. In *IFIP World IT Forum (WITFOR)*.(pp. 43–48) Springer
18. Jazdi, N. (2014). Cyber physical systems in the context of Industry 4.0. In *IEEE International Conference on Automation, Quality and Testing, Robotics* (pp. 1–4). IEEE Publications.
19. Tan, Y., Goddard, S., & Pérez, L. C. (2008). A prototype architecture for cyber-physical systems. *ACM SIGBED Review*, 5(1), 1–2. https://doi.org/10.1145/1366283.1366309.
20. Han, S., Xie, M., Chen, H.-H., & Ling, Y. (2014). Intrusion detection in cyber-physical systems: Techniques and challenges. *IEEE Systems Journal*, 8(4), 1052–1062. https://doi.org/10.1109/JSYST.2013.2257594.
21. Lin, J., Yu, W., Zhang, N., Yang, X., Zhang, H., & Zhao, W. (2017). A survey on internet of things: Architecture, enabling technologies, security and privacy, and applications. *IEEE Internet of Things Journal*, 4(5), 1125–1142. https://doi.org/10.1109/JIOT.2017.2683200.
22. Guan, Y., & Ge, X. (2017). Distributed attack detection and secure estimation of networked cyber-physical systems against false data injection attacks and jamming attacks. *IEEE Transactions on Signal and Information Processing Over Networks*, 4(1), 48–59. https://doi.org/10.1109/TSIPN.2017.2749959.
23. Peng, Y., Lu, T., Liu, J., Gao, Y., Guo, X., & Xie, F. (2013). Cyber-physical system risk assessment ninth international conference on intelligent information hiding and multimedia signal processing, 2013, p. 169.
24. Lu, T., Xu, B., Guo, X., Zhao, L., & Xie, F. (2013). *A New Multilevel Framework for Cyber-Physical System Security* (pp. 2–3). Springer.
25. Konstantinou, C., Maniatakos, M., Saqib, F., Hu, S., Plusquellic, J., & Jin, Y. (2015). Cyber-physical systems: A security perspective. In *20th IEEE Europa Test Symposium* (pp. 1–8). IEEE European Test Symposium.
26. Raza, S. (2013). *Lightweight Security Solutions for the Internet of Things, Mälardalen University Press Dissertations*. Mälardalen University.
27. Wang, E. K., Ye, Y., Xu, X., Yiu, S. M., Hui, L. C. K., & Chow, K. P. (2010). Security issues and challenges for cyber physical system. In *IEEE/ACM International Conference on Green Computing and Communications and 2010 IEEE/ACM International Conference on Cyber, Physical and Social Computing* (pp. 733–738). Springer.
28. Weiss, J. (2010). *Control System Cyber Vulnerabilities and Potential Mitigation of Risk for Utilities, White Pap*. Juniper Networks, Inc.
29. Hu, W., Oberg, J., Barrientos, J., Mu, D., & Kastner, R. (2013). Expanding gate level information flow tracking for multilevel security. *IEEE Embedded Systems Letters*, 5(2), 25–28.
30. Jing, Q., Vasilakos, A. V., Wan, J., Lu, J., & Qiu, D. (2014). Security of the internet of things: Perspectives and challenges. *Wireless Networks*, 20(8), 2481–2501. https://doi.org/10.1007/s11276-014-0761-7.
31. Stankovic, J. A. (2014). Research directions for the internet of things. *IEEE Internet of Things Journal*, 1(1), 3–9. https://doi.org/10.1109/JIOT.2014.2312291.
32. Tawalbeh, L. A., Mowafi, M., & Aljoby, W. (2013). Use of elliptic curve cryptography for multimedia encryption. *IET Information Security*, 7(2), 67–74. https://doi.org/10.1049/iet-ifs.2012.0147.
33. Rungger, M., & Tabuada, P. (2013). A notion of robustness for cyber-physical systems. *IEEE Transactions on Automatic Control*, 61(8), 2108–2123. https://doi.org/10.1109/TAC.2015.2492438.
34. Lo'ai, A. T., Tawalbeh, L. A., Mehmood, R., Benkhlifa, E., & Song, H. (2016). Mobile cloud computing model and big data analysis for healthcare applications. *IEEE Access*, 4, 6171–6180. https://doi.org/10.1109/ACCESS.2016.2613278.

35. Tawalbeh, A., Haddad, Y., Khamis, O., Benkhelifa, E., Jararweh, Y., & AlDosari, F. (2016). Efficient and secure software-defined mobile cloud computing infrastructure. *International Journal of High Performance Computing and Networking*, 9(4), 328–341. https://doi.org/10.1504/ IJHPCN.2016.077825.

36. Kocher, P. C. (1996). Timing attacks on implementations of Diffie-Hellman, RSA, DSS, and other systems. In *Lecture Notes in Computer Science. Proceedings of the Cryptology, Santa Barbara* (pp. 104–113). https://doi.org/10.1007/3-540-68697-5_9.

37. Lu, T., Lin, J., Zhao, L., Li, Y., & Peng, Y. (2015). A security architecture in cyber-physical systems: Security theories, analysis, simulation and application fields. *International Journal of Security and its Applications*, 9(7), 1–16. https://doi.org/10.14257/ijsia.2015.9.7.01.

38. Zhao, K., & Ge, L. (2013). A survey on the internet of things security. In *Proceedings of the 2013 Ninth International Conference on Computational Intelligence and Security* (pp. 663–667). ACM.

39. Shafi, Q. (2012). Cyber physical systems security: A brief survey. In 2012 12th International Conference on Computational Science and Its Applications (pp. 146–150).IEEE.

40. Bhattacharya, R. (2013). A comparative study of physical attacks on wireless sensor networks. *International Journal of Research in Engineering and Technology*, 5, 72–74.

41. Alvaro, C., Amin, S., Sinopoli, B., Giani, A., Perrig, A., & Sastry, S. (2009). Challenges for securing cyber physical systems. *Workshop on Future Directions in Cyber-physical Systems Security*, 5, 45–50.

42. Mo, Y., & Sinopoli, B. (2012). Integrity attacks on cyber-physical systems. In *Proceedings of the 1st International Conference on High Confidence Networked Systems* ACM (pp. 47–54).

43. Gaddam, N., Kumar, G. S. A., & Somani, A. K. (2008). Securing physical processes against cyber attacks in cyber-physical systems. In *High Confidence Transportation Cyber-Physical Systems: Automotive, Aviation, and Rail, Washington DC* (pp. 2–4). https://people.cs.ksu.edu/ ~danielwang/Investigation/CPS_Security_threat/ArunSomani_20081024_emailsubmission.pdf

44. Gamage, T. T., Roth, T. P., & McMillin, B. M. (2011). Confidentiality preserving security properties for cyber-physical systems. In *35th IEEE Annual Computer Software and Applications Conference* IEEE (pp. 28–37).

45. Madden, J., McMillin, B., & Sinha, A. (2010). Environmental obfuscation of a cyber physical system—Vehicle example. In *Workshop on 34th Annual IEEE Computer Software and Applications Conference* IEEE (pp. 176–181).

46. Venkatasubramanian, K. K. (2009). 'Security solutions for cyber-physical systems,' A [Dissertation] presented in partial fulfillment of the requirements for the degree doctor of philosophy.

47. Parvin, S., Hussain, F. K., Hussain, O. K., Thein, T., & Park, J. S. (2013). Multi-cyber framework for availability enhancement of cyber physical systems. *Computing*, 95(10–11), 927–948. https:// doi.org/10.1007/s00607-012-0227-7.

48. Seering, J., Flores, J. P., Savage, S., & Hammer, J. (2018). The social roles of bots: Evaluating impact of bots on discussions in online communities. *Proceedings of the ACM on Human-Computer Interaction*, 2(CSCW), 1–29. https://doi.org/10.1145/3274426.

49. Srivastava, A., Morris, T., Ernster, T., Vellaithurai, C., Pan, S., & Adhikari, U. (2013). Modeling cyber-physical vulnerability of the smart grid with incomplete information. *IEEE Transactions on Smart Grid*, 4(1), 235–244. https://doi.org/10.1109/TSG.2012.2232318.

50. Hawrylak, P. J., Haney, M., & Papa, M. H. (2012). Using hybrid attack graphs to model cyber-physical attacks in the Smart Grid. In *5th International Symposium on Resilient Control Systems* IEEE (pp. 161–164).

51. Pasqualetti, F., Dorfler, F., & Bullo, F. (2011). Cyber-physical attacks in power networks: Models, fundamental limitations and monitor design. In *50th IEEE Conference on Decision and Control and European Control Conference* IEEE (pp. 2195–2201).

52. Kuhl, M. E., Kistner, J., Antini, K. C., & Sudit, M. (2007). Cyber attack modeling and simulation for network security analysis simulation conference. In *2007 Winter Simulation Conference* IEEE (pp. 1180–1188).

53. Seiger, R., Keller, C., Niebling, F., & Schlegel, T. (2015). Modelling complex and flexible processes for smart cyber-physical environments. *Journal of Computational Science*, 10, 137–148. https://doi.org/10.1016/j.jocs.2014.07.001.

54. Kroiß, C., & Bureš, T. (2016). Logic-based modeling of information transfer in cyber–physical multi-agent systems. *Future Generation Computer Systems*, 56, 124–139. https://doi.org/10.1016/j .future.2015.09.013.

55. Khaitan, S. K., & McCalley, J. D. (2015). Design techniques and applications of cyberphysical systems: A survey. *IEEE Systems Journal*, 9(2), 350–365. https://doi.org/10.1109/JSYST.2014 .2322503.

56. Petnga, L., & Austin, M. (2016). An ontological framework for knowledge modeling and decision support in cyber-physical systems. *Advanced Engineering Informatics*, 30(1), 77–94. https:// doi.org/10.1016/j.aei.2015.12.003.

57. Kelly (Letcher), R. A., Jakeman, A. J., Barreteau, O., Borsuk, M. E., ElSawah, S., Hamilton, S. H., Henriksen, H. J., Kuikka, S., Maier, H. R., Rizzoli, A. E., van Delden, H., & Voinov, A. A. (2013). Selecting among five common modelling approaches for integrated environmental assessment and management. *Environmental Modelling and Software*, 47, 159–181. https://doi.org/10 .1016/j.envsoft.2013.05.005.

58. Strasser, U., Vilsmaier, U., Prettenhaler, F., Marke, T., Steiger, R., Damm, A., Hanzer, F., Wilcke, R. A. I., & Stötter, J. (2014). Coupled component modelling for inter- and transdisciplinary climate change impact research: Dimensions of integration and examples of interface design. *Environmental Modelling and Software*, 60, 180–187. https://doi.org/10.1016/j .envsoft.2014.06.014.

59. Zhu, M., & Martínez, S. (2011). Stackelberg-game analysis of correlated attacks in cyber-physical systems. In *Proceedings of the American Control Conference, San Francisco, United States* USA (pp. 4063–4068).

60. Pasqualetti, F., Dorfler, F., & Bullo, F. (2013). Attack detection and identification in cyber-physical systems. *IEEE Transactions on Automatic Control*, 58(11), 2715–2729. https://doi.org/10.1109/ TAC.2013.2266831.

61. Slay, J., & Miller, M. (2007). Lessons learned from the Maroochy water breach. *Critical Infrastructure Protection*, 253, 73–82.

62. Mitchell, R., & Chen, I. R. (2011). A hierarchical performance model for intrusion detection in cyber-physical systems. In *Wireless Communications and Networking Conference (WCNC)* IEEE (pp. 2095–2100).

63. Mitchell, R., & Chen, I. R. (2014). Behavior rule specification-based intrusion detection for safety critical medical cyber physical systems. *IEEE Transactions on Dependable and Secure Computing*, 12(1), 16–30. https://doi.org/10.1109/TDSC.2014.2312327.

64. Pan, S., Morris, T. H., Adhikari, U., & Madani, V. (2011). *Causal Event Graphs Cyber-Physical System Intrusion Detection System Mississippi State University*. USA.

65. Dorfler, F., Pasqualetti, F., & Bullo, F. (2011). Distributed detection of cyber-physical attacks in power networks: A waveform relaxation approach. In *49th Annual Allerton Conference on Communication, Control, and Computing (Allerton)* Allerton (pp. 1486–1491).

66. Mitchell, R., & Chen, I. R. (2014). A survey of intrusion detection techniques for cyber-physical systems. *ACM Computing Surveys*, 46(4), 1–29. https://doi.org/10.1145/2542049.

67. Mitchell, R., & Chen, I. (2015). Behavior rule specification-based intrusion detection for safety critical medical cyber physical systems. *IEEE Transactions on Dependable and Secure Computing*, 12(1), 16–30.

68. Alcaraz, C., Cazorla, L., & Fernandez, G. (2015). Context-awareness using anomaly-based detectors for smart grid domains. In *Lecture Notes in Computer Science International Conference on Risks and Security of Internet and Systems* (pp. 17–34). Springer. https://doi.org/10.1007/978-3 -319-17127-2_2.

69. Abbas, W., Laszka, A., Vorobeychik, Y. et al. (2015). Scheduling intrusion detection systems in resource-bounded cyber- physical systems. In *Proceedings of the 1st ACM Workshop on Cyber-Physical Systems-Security and/or Privacy. Academic Medicine*, 55–66.

70. Naghnaeian, M., Hirzallah, N., & Voulgaris, P. G. (2015). Dual rate control for security in cyber-physical systems. In *54th IEEE Conference on Decision and Control (CDC)* IEEE (pp. 14145–11420).
71. Ivanov, R., Pajic, M., & Lee, I. (2016). Attack-resilient sensor fusion for safety-critical cyber-physical systems. *ACM Transactions on Embedded Computing Systems*, 15(1), 1–24. https://doi.org /10.1145/2847418.
72. Zimmer, C., Bhat, B., Mueller, F. et al. (2010). Time-based intrusion detection in cyber-physical systems. In *Proceedings of the 1st ACM/IEEE International Conference on Cyber-Physical Systems. Academic Medicine*, 109–118.
73. American Gas Association. (2005). Technical report. "Cryptographic protection of SCADA communications part 1: Background, policies and test plan". AGA Report.
74. Wan, K., & Alagar, V. (2013). Context-aware security solutions for cyber-physical systems. *Lecture Notes of the Institute for Computer Sciences, Social-Informatics and Telecommunications Engineering*, 109, 18–29. https://doi.org/10.1007/978-3-642-36642-0_3.
75. Sridhar, S., Hahn, A., & Govindarasu, M. (2012). Cyber physical system security for the electric power grid. *Proceedings of the IEEE*, 100(1), 210–224. https://doi.org/10.1109/JPROC.2011 .2165269.
76. Weiss, J. M. (2007). 'Control systems cyber security—The need for appropriate regulations to assure cyber security of the electric grid', Testimony [Report]. To Homeland Security's Subcommittee on Emerging Threats, Cyber-Security, and Science and Technology.
77. Chakraborty, N. (2013). Intrusion detection system and intrusion prevention system: A comparative study. *International Journal of Computing and Business Research (IJCBR)*, 4(6166), 2229.
78. Almohri, H., Cheng, L., Yao, D., & Alemzadeh, H. (2017). On threat modeling and mitigation of medical cyber-physical systems. In IEEE/ACM International Conference on Connected Health: Applications, Systems and Engineering Technologies (CHASE) IEEE and ACM (pp. 114–119).
79. Shu, X., Yao, D., & Ramakrishnan, N. (2015). Unearthing stealthy program attacks buried in extremely long execution paths. In *Proceedings of the 22nd ACM SIGSAC Conference on Computer and Communications Security. Academic Medicine*, 401–413.
80. Xu, K., Tian, K., Yao, D., & Ryder, B. G. (2016). A sharper sense of self: Probabilistic reasoning of program behaviors for anomaly detection with context sensitivity. In *46th Annual IEEE/IFIP International Conference on Dependable Systems and Networks (DSN) IEEE* IEEE (pp. 467–478).
81. Urbina, D. I., Giraldo, J. A., Cardenas, A. A., Tippenhauer, N. O., Valente, J., Faisal, M., Ruths, J., Candell, R., & Sandberg, H. (2016). Limiting the impact of stealthy attacks on industrial control systems. In *Proceedings of the 2016 ACM SIGSAC Conference on Computer and Communications Security. Academic Medicine*, 1092–1105.
82. Jiang, N., Lin, H., Yin, Z., & Xi, C. (2017). Research of paired industrial firewalls in defense-in-depth architecture of integrated manufacturing or production system. In 2017 IEEE International Conference on Information and Automation (ICIA) IEEE (pp. 523–526).
83. Ghosh, T., Sarkar, D., Sharma, T., Desai, A., & Bali, R. (2016). Real time failure prediction of load balancers and firewalls. In 2016 IEEE International Conference on Internet of Things (iThings) and IEEE Green Computing and Communications (GreenCom) and IEEE Cyber, Physical and Social Computing (CPSCom) and IEEE Smart Data (SmartData) IEEE (pp. 822–827).
84. Cohen, F. (2006). The use of deception techniques: Honeypots and decoys. *Handbook of Information Security*, 3(1), 646–655.
85. Irvene, C., Formby, D., Litchfield, S., & Beyah, R. (2017). Honeybot: A honeypot for robotic systems. *Proceedings of the IEEE*, 106(1), 61–70. https://doi.org/10.1109/JPROC.2017.2748421.
86. Seeber, S., Sehgal, A., Stelte, B., Rodosek, G. D., & Schönwälder, J. (2013). *Towards A Trust Computing Architecture for RPL in Cyber Physical Systems*. In *Proceedings of the 9th International Conference on Network and Service Management (CNSM 2013)* (pp. 134–137). International Federation for Information Processing/IEEE International Conference on Network and Service Management (CNSM 2013).
87. Haehner, J., Rudolph, S., Tomforde, S., & Fisch, D. (2013). *A Concept for Securing Cyber-Physical Systems with Organic Computing Techniques* IEEE (pp. 1–13).

88. McMahan, H. B., Moore, E., Ramage, D. et al. (2017). Communication-efficient learning of deep networks from decentralized data. In *International Conference on Artificial Intelligence and Statistics* IEEE (pp. 1273–1282).

89. McAfee special report: How collaboration can optimize security operations. The new secret weapon against advanced threats. (2016). https://abyteofcyber.com/DOCS/rp-soc-collaboration-advanced-threats.pdf.

10

Cyber Physical System Approach for Smart Farming and Challenges in Adoption

Rajesh Kumar, Amit Pandey, and Kuldeep Singh Kaswan

CONTENTS

10.1 Introduction

Among the most significant difficulties highlighted for the twenty-first century are:

- Scarcity of resources – environmental challenges
- An aging population
- Rapid worldwide population expansion and widespread urbanization
- Globalization and increasing mobility
- Increased competitive pressure
- High dynamism in technological development
- Increased middleware

All of these have significant consequences for the most critical aim of our century: ensuring the correct framework for feeding and supporting the whole world's population over an indefinite "time horizon" [1].

> The quality and diversity of agriculture items as crucial characteristics of the population's nutritional value have long been a central objective of both experts and, particularly,

DOI: 10.1201/9781003220664-10

administrations. On the other hand, there is a growing recognition that, in the case of increased demography, not only are sustainable farming surfaces constricted, but the impact of climate change on crop production may be so powerful that, in the very worst sort of situation, the important environmental equivocation that allows all above congregation to exist is likely to be destroyed.

Already, technology has had an effect "on the investment efficiency of agricultural production: automation, artificial fertilizers, pharmaceutical pesticide application, and information and technology have long proven their usefulness in raising the quantity of food while requiring less human labor" [2].

Innovative products (biotechnologies, planning, and control) have also been introduced to reduce negative environmental consequences and improve the efficiency of operations of the agricultural sector. However, technology advancements have been employed in the pato assist one or more components of the farm sector, without a comprehensive perspective of their interdependencies, rely without taking into account the feedback mechanism, which includes the long-term risk associated.

"Agriculture's development, particularly in the last fifty years, has favored specialization food and fiber productivity rose due to new technologies (mechanized farming, automating, chemistry, and microbiology) and regulations favor maximizing output in response to market forces. Because of these advances, fewer producers with lower labor needs could provide the bulk of food and fiber in the most industrialized economies" [3].

"While these modifications have had numerous good benefits and decreased many dangers in farming," as stated in [1] "there have also been major expenses." Topsoil degradation, groundwater pollution, the demise of agricultural producers, persistent disregard of farm laborer social and studying circumstances, rising production costs, and the breakdown of the standard of living in rural areas are noteworthy among these [4].

During the last two decades, a significant movement has developed to challenge the agribusiness leadership's role in encouraging practices that contribute to these socioeconomic problems. Today, the sustainable agricultural movement is gaining traction and acceptability in mainstream agriculture. Sustainable agriculture not only addresses numerous environmental issues, but it also provides creative and commercially viable options for producers, laborers, consumers, legislators, and so many others across the food chain. The referencing above is significant of a growing phenomenon in which the farmland entrepreneurship model is shifting toward greater complexity, with behavior emerging from the communication networks and coexistence of multiple portions that are independent of the legislature and knowledgeable, through promoting best practices, of achieving a fully functioning optimal solution, sharing "costs, and especially reducing the environmental footprint. The environment is especially considered in this model, both as a specialized efficiency requirement and as an explicit feature of the model" [5].

System dynamics, "ICT, and" information sharing now can enable the development of such a framework by utilizing one of the most recent and demanding paradigms – cyber physical systems.

This chapter seeks to give a systematic agricultural approach. Enterprise is described as a dynamic system whose emergence behavior may be altered by changes in the external environment.

The following part will introduce the notion of agricultural production enterprise used in this work and the essential criteria for its operation.

A model of this firm will be given, emphasizing the complicated flow of information and goods and their linkage.

The notion of cyber physical systems will be briefly described as solutions and technologies for an intelligence agriculture organization central controller, as an interconnected multi-agent wireless mesh network. "Such design may be able to facilitate a lean migration from current industrially focused agriculture to sustainable farming in short, realistic phases [6]."

10.2 Cyber-Physical Farming Platform

This work integrates traditional agricultural practices with modern cyber-enabled technologies to build adaptable SeDS diagnostic systems and assist agriculture. This study employs both qualitative and quantitative approaches to overcome the absence of needed data as well as related inadequacies in cyber-physical agricultural systems. As mentioned in this part, the integrative technique in this work evaluates multiple factors (e.g., soil physical, chemical, and biological characteristics) and variables (e.g., crop yield and quality) under real-world settings to enhance the dimensionality of CPS and data-driven selection. The "bottom-up" method to describing and forecasting local soil and environmental conditions suggested herein indicates a possible paradigm change from existing tactics that seek to improve production exclusively by relying on general, empirical connections between fertilizers rate and agricultural yield [7]."

10.2.1 Improving Crop Productivity

The sensitivity and specificity of geostationary pictures "are in the neighborhood of feet per pixels (a field-of-view of 10–50 km), but there is also lacking imaging data for land regions under cloud covering under overcast meteorological conditions. Despite the disadvantages highlighted, drones constitute a desirable decision for a visual experience of earth occurrences and cultivated lands since they are inexpensive to purchase for on-demand instant scanning and picture processing [8]. Drone imaging has a positioning accuracy ranging from 0.25 to 4 inches per pixel (a field-of-view of 10–500 m). As a result of the tremendous temporal granularity given by drone photos, soil health and crop productivity may be assessed more often in developing field locations. This capability is highly beneficial to farmers since it simplifies the impact evaluation. Drone-delivered hyperspectral detectors are used to study the effects of a wide range of fertilizer types and application rates. These webcams are becoming more popular in crop monitoring. Compared with conventional display information, which provides image feature pathways connected with the electromagnetic radiation in the visible spectrum's Red, Green, and Blue (RGB) light-specific wavelength, infrared imaging cameras provide multiple channels of visual energy in the near-infrared band of the electromagnetic spectrum. As a result, these cameras gather crop intelligence that farmers cannot see or receive with traditional RGB cameras [9]. Because it is connected to pigmentation, it is instrumental in assessing crop performance (natural coloring of plant tissue). Precisely, vegetation having low chlorophyll levels (as the primary photosynthetic pigment in plants) reflects less light in the near-infrared spectrum than young seedlings. Similarly, decreased leaf hydration causes a shift in the transmitted wavelength content used to detect water scarcity in crops. Furthermore, increased drone data may be utilized to extract important crop properties, such as leaf shape and color patterns that are not recorded in low-resolution aerial photographs. Certain phenotypic (for example, metabolic and molecular qualities) are distinguished by variations in the morphology of the leaves without changing the crop's spectral content. In addition, leaf coloration characteristics can determine sick plants from nutritional ones. This work investigates the use of fieldwork mappings for numerous regularly used switch virtual interfaces (SVIs), as well as leaf shape-color metadata, in the building of business intelligence algorithms for assessing the influence of nutrients on organically crop output [10]."

10.2.2 Sensing Soil Capability to Improve Crops

"This chapter seeks to deliver efficient and accessible decision support and services by developing data-driven algorithms for diagnosing soil health issues and detecting temporal variations [11]. Accurate soil crop (SC) health monitoring is critical for increasing SC durability and supporting agricultural operations. Soil health assessment models use deep data-driven models learned from gathered data to depict intricate interactions among degraded soils. SC health monitoring gives near-real-time data (data communication delays) that enables farmers to make informed decisions. Soil health is an integrating attribute that deals with soil geochemical, microbiological, and physical qualities and reflects the fertility of the soil to support agriculture productivity and ecological services consistently [12]. It depends on the preservation of four essential operations: carbon alterations, nutrient cycles, soil characteristics, and pest and disease control; however, due to the very complicated interconnections of the processes, diagnosing soil health by evaluating each function independently is inadequate. The data-driven models developed in this study soil improvement of crop yields by utilizing multi-source acquired data (detailed in Phase I) and exploiting Deep Neural Networks' high hierarchical data representational capacity (DNNs). The obtained data and observations are divided into several groups based on the soil-crop function. DNNs offer an automatic representation-learning strategy for raw data, allowing models to conform to complicated relationships. DNN models that have been created learn the association between soil/crop data and health/performance metrics [13]."

10.2.3 Crop Performance and Soil Improvement Analysis

"In conjunction with Phases 1 and 2, this process intends to strengthen a cyber-physical agricultural framework for real-time measurement and reporting of numerous environmental variables (e.g., NPK values) as well as recognizing conservation agriculture combinations to disclose better overall crop performance through techno-economic and social-economic research findings [14]. A field-deployable identity soil surveillance data collecting device is presented for obtaining real-time and regular observations of several soil quality variables of interest. The sensory analysis system comprises numerous basic components, including various soil-making data-driven decisions (such as pH and NPK), a sensor aggregating and management boards, a communication module, and a data-driven decision-making block as a communications network. The suggested cyber-physical infrastructure for the communication system and the interactions of the elements of the system are being watched. The sensing packages interface with the inbuilt microcontroller and transmit data using the 3G/4G protocol via the transceiver. To extend the range of the sensor channels, a repeater station might be installed in the field under monitoring, which would receive data from numerous sensor platforms and transmit it to a field station for synchronizing. This data is kept in the cloud to allow for remote collection and management with stakeholders to enhance farming techniques instantaneously and daily [15]."

10.3 Challenging Factor in Agricultural

This chapter analyzes sustainable management as an expression of the notion that we must address the requirements of the present without exacerbating existing issues or jeopardizing existing resources [16].

As a result, sustainable agriculture has three primary connected goals: preserving/improving ecological integrity, preserving/ensuring (an acceptable degree of) net cash flow, and promoting social and economic fairness.

In the following sections, we will go through the components involved in assuring goal attainment about the global issues outlined in the introduction.

Scarce resources – environmental pressures: Water, land, air, and electricity are the primary natural and ecologically conscious resources implicated in the agricultural industry. On the other hand, being a comprehensive supply from the atmosphere, the weather may be viewed as a deciding element in fertilizer application and a significant stochastic disruption [17].

Water quality is one of the most critical resources for agricultural output. It is estimated that 2/3 of the Canadian population suffered from lack of it in some form or another. Not only does agriculture require water, but the widespread use of synthetic chemicals, which could improve efficiency with less effort than other alternative solutions, can affect water and, as a result, either start reducing this commodity or imply, over time, increased costs for its detoxification. Water management, encompassing water transport and storage, and sophisticated irrigation schemes with continuous operation and assessment of local soil moisture via networked sensor systems and controller design policies should become a key issue in agricultural enterprises. Reusing and purifying water when measuring the productivity of the agricultural food manufacturing operation, expenses should be addressed [18]."

Floods and water shortages should be considered not only at the water distribution system (by adequate resource stockpiling and preservation of natural resources, as well as a good design and implementation of irrigated agriculture), but also at an elevated, worldwide platform of operational management, through good speciation use, fertilizer application, and increased automation operations for water fault detection.

Soil is the key resource for agricultural, and it is not only naturally restricted, but it is also shrinking in the surface area owing to growing populations [19]."

In reality, it may be argued that productive agricultural landscapes are shrinking in size, not simply due to the substantial growth of agriculture but densely populated locations and due to excessive soil extraction and phenomena caused by climate "risk factors such as desertification."

Soil erosion is a significant danger to our capacity to generate enough food. Soil management encompasses various measures designed to maintain the soil in places, such as reducing or even eliminating tillage, using irrigation scheduling management to decrease runoff, and keeping the ground covered with vegetation or mulch. Soil quality not only has a significant impact on food production, dictating its structure and expenses but it may also be influenced by cultures, serving as both an input and a production for the agricultural sector as a whole [20].

We are taking the earth as a fundamental and fundamental resource, densely populated locations, and due to excessive soil extraction and phenomena caused by the climate risk factors such as desertification.

Soil erosion is a significant danger to our capacity to generate enough food. Soil management encompasses various measures designed to maintain the soil in places, such as reducing or even eliminating tillage, using irrigation scheduling management to decrease runoff, and keeping the ground covered with vegetation or mulch. Soil quality not only has a significant impact on food production, dictating its structure and expenses but it may also be influenced by cultures, serving as both an input and a production for the agricultural sector as a whole [21].

Considering Earth itself to be a primary resource, renewable energy, especially hydrocarbons, it would be economically unsustainable to quickly abandon or even reduce the works (mostly mechanical) that depend on these nonrenewable energy sources. On the other hand, sustainable farming practices need to look for extra reliable and preferably electric resources (like energy obtained from organic waste); until then, excellent management of the activities involved in agricultural production could decrease energy usage.

- Population aging: An older society means reducing skilled workers, which could be made obsolete, but at a penalty in terms of energy assets. Furthermore, this population sector, in particular, is driving up the production of organic, healthful products [22].

- Population and rising and widespread urbanization: This entails a constant loss of resources such as soil and water and a strain on the economic process of productivity, as food and fiber output should increase as resources are depleted. Furthermore, an unintended consequence of urbanization is the degradation of air quality. In addition to having an immediate impact on the urban microclimate, air pollution has the potential to trigger far-reaching weather patterns over time [23]."

- Globalization and greater mobility: Assurance and the potential to alter both the distribution and distribution patterns of agricultural goods. Canadian currency and financial pressure on the agribusiness company's present expenses and revenues; on the other hand, the growing susceptibility of the repercussions of disruptions in certain regions of the world to spread due to rising globalization [24].

- Increasing economic pressure: Profit drives any business, and its degree significantly impacts how innovators conduct their work. There is high competition between companies worldwide, resulting in high dynamism in client demands and revenue maximization. These developments may have a significant impact on agribusiness businesses.

- Highly technological advancement dimension: Managed to install mechanical and mechanized systems in agriculture, increasing production. Production, a reduction in the number of people employed, and more efficient soil utilization overall. In addition, advances in ICT have resulted in more significant usage of these techniques at every stage of agricultural production, laying the groundwork for a paradigm shift in which the notion of cyber physical systems (CPS) might find a wide range of applications [25]."

- Facilitating a Greater degree of communications: This factor may be a driver for the intensive farming organization, as it means knowledge that could be used to enhance functional departments and respond swiftly to environmental factors is gathered through sound surveillance systems while keeping all of the above variables in mind through adequate measuring devices. These sources include things like weather forecasting, transfer stock rates, and supply chain partners.

10.3.1 Agricultural Techniques

As mentioned in the preceding section, Figure 10.1 depicts a modular form of the supervised agricultural experience (SAE).

The firm is believed to have both vegetable and animal products, resulting in three critical subsystems:

- the land management subsystems, which encompasses all aspects of culture and pasture scheduling, fertilization, and soil dampness continuous monitoring, as well as natural conditions, weather forecasts, market predictions, business goals, and producing technologies and knowledge [26];

- The production planning sub-system deals with achieving goals specified by the soil conservation comment thread for crops; it considers elements such as seed production, work planning based on crop-specific characteristics and growing conditions, weather patterns, and plant population. The conservation farming component has similar objectives but is concerned with animals.

- The first step of this process is to ensure that the terrain has the best enslavement conditions possible, using different methods tailored to soil quality and snow conditions, weather predictions, and long- and medium-term crop and livestock planning to meet customer expectations and full compliance.

A sensing network must continuously monitor temperatures, moisture, and nutrient content in the soil to control and update data about soil quality via automated wagering systems and fertilizer distributors. These technologies lead to building automation at the physical level of agricultural production, allowing for consistent customer satisfaction, which is the second step. The essential characteristic is that the first integration is based on a system structure, which helps both the sensory system and the control network to be constructed step by step depending on the importance of desired functionalities and environmental conditions to be considered. Each type of organization and any predefined agriculture surface can be described as an agent whose attributes and operating rules

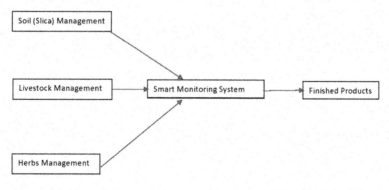

FIGURE 10.1
Smart agricultural monitoring system.

are stored in the world model and whose current state and evolution can be separately upgraded, as shown in the implementation approach [27].

As the network's dimensions grow, the agricultural company's complicated behavior will emerge with higher precision. The real benefit of this method is that humans may participate. As autonomous algorithms and knowledge consumers, the agent architecture provides sufficient expertise while also adding a lot of freedom.

The existence of a detection network – not only multisensory networks, but also more elaborate gadgets that can construe various intelligence gathered from channels such as sensors, living beings, and even news stations to recognize and even foresee pertinent changes whereby the farm could perhaps adjust – appears to be required to guarantee the flexibility of the producing to external conditions – weather, market, diseases [28].

In the last stage of deployment, agents work together to form a system that displays emergent proactive behavior as a result of linking its inherent structure (as supplied by the SAE). Stage I entails a production schedule (as completed by the AI system), while Stage II involves an evaluation of the external world's evolution (regarding the SAE) based on the sensing network and the perception of its knowledge via an effective knowledge administering and context successful track record of integrating.

In this model, two massively parallel ecosystems could be recognized: one for production, which is severely affected by climate, weather patterns, pests, and other factors, and another for product-market capitalization, which is influenced primarily by environmental factors with a highly dynamic system progression.

Intelligent process and reporting systems, which connect sensing, actuators, computers, and telecommunication technologies, could control the first one into a scalable ability to scale with reliable and adaptive governance.

The second ecosystem is susceptible to market-driven disruptions that are difficult to foresee and even appropriately detect.

SAE will have a comprehensive system monitoring and assessment, planning and control of crop development, and, last but not least, a production and distribution management system as an intelligent farm.

Due to its integration into a global information system, SAE will have access to services such as complex, large datasets containing a variety of crop selection options, correlated with specific soil characteristics, weather forecasts, and market developments, as well as the decision support that comes along with them, techniques for efficiently handling agricultural outputs and scoring the greatest possible price on the global market.

The goal of an irrigation canal with actual self and acclimation functionality at environment protection, snow conditions, and market mechanisms and instabilities in the general framework of an autonomous, conservation and economical, and socially responsible improvement is considered through appropriate machine collection, enhanced digital transformation, and efficient subjugation of multi-annual data sets in the decision-making process."

10.4 Agricultural Approach

As mentioned before, each component of an SAE paradigm can be thought of as a separate but collaborating individual. It has to be able to process information related to its major aims utilizing synaptic plasticity, evaluate its suitability with the stated goals along the course of achieving its objectives, deliberate on goal revisions, and act it toward the environment.

10.4.1 Agricultural Tools

Since the success of this module depends on the participation of these other entities, gathering data from a variety of sources belonging to them should be the top priority.

Complex behaviors might emerge from the hierarchical structure of these agents, which could simulate various aspects of SAE during various stages of development and deployment.

10.4.2 Cyber Physical Systems

"Cyber-physical systems (CPS) are designed systems that are developed from and rely on the symbiosis of computing and factors more specific," according to [3]. Developing CPS will be synchronized, decentralized, and networked, and they will need to be both resilient and adaptive. Today's new CPS will far outperform today's simple, intelligent systems in terms of features, flexibility, robustness, safety, reliability, and accessibility. CPS innovation will change how people engage with designed systems in the same way that the internet transformed how people interact with knowledge. In areas such as the electricity grid, transportation, architecture, medicine, and manufacture, modern digital computer systems will fuel competitiveness.

A cyber physical system (CPS) is a system that:"

- is strongly interwoven and non-separable from a task-scheduling processor and processes that take place from a behavioral standpoint;
- where the interplay of computing and communication components reveals functioning and critical system properties;
- in which machines, systems, equipment, and the surroundings in which they are embedded have physicochemical characteristics that interact with one another, use resources, and lead to significant system behavior.

These systems will alter our methods of dealing with the external surroundings in the same way that the internet has changed social communication.

Economic and social drivers, such as falling costs of computation, connections, and sensor systems; rising demands for energy-efficient use of extensive and complex infrastructure facilities; and conservation of natural for technological solutions that make efficient use of renewable resources and produce less waste all contribute to the CPS worldview. It has genuine goods with usable information embedded in them.

The fundamental economic motive for implementing CPS in every business and application is lowering computing, communications, and sensor costs. Computer and telecommunications have evolved into the "universal system engineer" that joins disparate systems. They make it possible to build CPS networks global or regional.

It is only reasonable to regard the intensive agriculture industry as a dynamic process, operating on the premise of autonomously and interacting agents – either human or non-human – behaviors, as previously indicated.

By integrating networking and computing infrastructure with manufacturing industries, promoting sustainable development, and related components, CPS enables a more accessible evolution of "intelligent" structures. The voice wants and knowledge of the "customers" will be integrated across the global supply chain in the upcoming generations of enterprises in the twenty-first century. It will be a results-driven company that responds swiftly to client needs while reducing resource consumption and enhancing environmental stability and economic attractiveness.

The CPS approach might already be deployed on the first stage of an SAE model, also incorporating the apparatus, equipment, and devices that offer soil treatments, thereby applying control rules and minimizing the effects of perturbations that might act at this level (drought, rain excess, pests etc.).

CPS necessitates more complex surveillance, forecasting, and understanding of associated data and characteristics on the second floor, based on information fusion and organizational learning. Depending on the quality, productivity, and environmental needs, such agricultural systems and approaches form the paradigms of a critical component of success.

Gradually expanding operational diversification and mechanization level, such a system might be implemented, guaranteeing a lean migration from a homogenous specialized income-producing property to a flexible, versatile, and responsive upgraded one.

10.5 Future Studies

Improved monitoring and complete evaluations can also provide appropriate soil/crop performance evaluation; however, current methods are prohibitively expensive and don't even provide real-time data. The following are some possible future research directions:

Result measurements for soil/crop evaluation and SDSS observation methodologies are being developed.

- Creation of a low-cost analysis tool that allows landowners to examine their farms' soil/crop health quickly.
- A comparative of crop yield and phenotypic expression in fields altered with organic fertilizer with areas handled with livestock wastes (e.g., bovine manure) or chemical fertilizers.
- Exploration mineralization rates are high enough to support agricultural production throughout the planting period, but they drop once the fallow season begins.
- Exploration of nutrients in contrast to the pulses of nutrients that soils receive following more contemporary agricultural practices.
- Exploration of biannual measurements of soil bulk density, carbon content, heavy metals, macroinvertebrates, and microbes."

10.6 Conclusions

Agricultural activity should really be re-designed as a sustainable food supply system for a continually growing populace in the framework of relevant climatic changes, extensive soil degradation, and reductions in drinking drinkable water sources. The rational utilization of mineral wealth (soil, water, and energy) in the context of global sustainable development has already been recognized as necessary for the planet's and its inhabitants' health. Real progress in industrial automation, biology, and "microbiology" has provided professionals with improved technological and organizational methods for producing high-quality food through the agricultural field. Significant outcomes from ICT made an application in the agricultural sector ensure more excellent production performance and accuracy. As a result, additional value is created in a field that is critical to humanity's survival. As an effectual agroecosystem, the intellectual farm is frequently mentioned, representing the effort to connect this industry in the digital reality of calculation and information exchange, allowing the Intelligent Cyber-Enterprise principle to be implemented in the field through the Availability Enterprise. As an integrated intelligent system, tomorrow's farming will allow for effective soil exploitation, environmental conservation, and, not least, the provision of ecologically cheap nutrients, all of which will have a significant impact on people's quality of life.

References

1. Hamuda E, Glavin M, Jones E (2016) A survey of image processing techniques for plant extraction and segmentation in the field. *Computers and Electronics in Agriculture* 125: 184–199.
2. Hersh B, Mirkouei A, Sessions J, Rezaie B, You Y (2019) A review and future directions on enhancing sustainability benefits across food-energy-water systems: The potential role of biochar-derived products. *AIMS Environmental Science* 6(5): 379–416.
3. Brenchley R et al. (2012) Analysis of the bread wheat genome using whole-genome shotgun sequencing. *Nature* 491(7426): 705–710.
4. Vogl GW, Weiss BA, Helu M (2016) A review of diagnostic and prognostic capabilities and best practices for manufacturing. *Journal of Intelligent Manufacturing* 30(1): 1–17.
5. Mirkouei A (2019) *Cyber-Physical Real-Time Monitoring and Controlling Biomass-Based Energy Production. Emerging Frontiers in Industrial and Systems Engineering: Growing Research and Practice.* Taylor & Francis, USA.
6. Elmore AJ, Stylinski CD, Pradhan K (2016) Synergistic use of citizen science and remote sensing for forest tree phenology continental-scale measurements. *Remote Sensing* 8(6): 502.
7. Candiago S, Remondino F, Del Giglio M, Dubbini M, Gattelli M (2015) Evaluating multispectral images and vegetation indices for precision farming applications from UAV images. *Remote Sensing* 7(4): 4026–4047.
8. Lottes P, Khanna R, Pfeifer J, Siegwart R, Stachniss C (2017) UAV-based crop and weed classification for smart farming. *Robotics and Automation (ICRA), IEEE International Conference on, IEEE.* Singapore.
9. Delgado-Baquerizo M et al. (2013) Decoupling soil nutrient cycles as a function of aridity in global drylands. *Nature* 502(7473): 672–676.
10. Panke-Buisse K, Poole AC, Goodrich JK, Ley RE, Kao-Kniffin J (2015) Selection on soil microbiomes reveals reproducible impacts on plant function. *The ISME Journal* 9(4): 980–989.
11. Challinor AJ et al. (2014) A meta-analysis of crop yield under climate change and adaptation. *Nature Climate Change* 4(4): 287–291.
12. Kaswan KS, Dhatterwal JS, Baliyan A, Jain V (2021). "Special sensors for autonomous navigation systems in crops investigation system" in the book title "virtual and augmented reality for automobile industry: Innovation vision and applications" published in springer series: *Studies in systems, decision and control,* January, 2022. Print ISBN 978-3-030-94101-7, Online ISBN 978-3-030-94102-4, https://doi.org/10.1007/978-3-030-94102-4_4.
13. Ray DK, Ramankutty N, Mueller ND, West PC, Foley JA (2012) Recent crop yield growth and stagnation patterns. *Nature Communications* 3: 1293.
14. Bradford JB et al. (2017) Future soil moisture and temperature extremes imply expanding suitability for rainfed agriculture in temperate drylands. *Scientific Reports* 7(1): 12923.
15. Wolfe ML et al. (2016) Engineering solutions for food-energy-water systems: It is more than engineering. *Journal of Environmental Studies and Sciences* 6(1): 172–182.

16. Hansen S, Mirkouei A, Xian M (2019) Cyber-physical control and optimization for biofuel 4.0. In Romeijn HE, Schaefer A, Thomas R (eds.) *Proceedings of the 2019 IISE Annual Conference.* Florida, USA, 75–85.

17. Dhatterwal JS, Kaswan KS, Pandey A (2022) "Implementation and deployment of 5G-drone setups", book entitled "internet of drones: Opportunities and challenges". In Apple Academic Press (AAP), Canada, Published Hard ISBN: 9781774639856.

18. Liu Z, Peng Y, Wang B, Yao S, Liu Z (2017) Review on cyber-physical systems. *IEEE/CAA Journal of Automatica Sinica* 4(1): 27–40.

19. Monostori L et al. (2016) Cyber-physical systems in manufacturing. *CIRP Annals-Manufacturing Technology* 65(2): 621–641.

20. Lee EA (2015) The past, present and future of cyber-physical systems: A focus on models. *Sensors (Basel)* 15(3): 4837–4869.

21. Gao R et al. (2015) Cloud-enabled prognosis for manufacturing. *CIRP Annals-Manufacturing Technology* 64(2): 749–772.

22. Hansen S, Mirkouei A (2018) *Past Infrastructures and Future Machine Intelligence (MI) for Biofuel Production: A Review and mi-Based Framework.* American Society of Mechanical Engineers, New York.

23. Lee J, Bagheri B, Kao H-A (2015) A cyber-physical systems architecture for industry 4.0-based manufacturing systems. *Manufacturing Letters* 3: 18–23.

24. Dhatterwal JS, Kaswan KS, Jaglan V, Vij A (2022) "Machine learning and deep learning algorithms for IoD". In book entitled "internet of drones: Opportunities and challenges" in "Apple Academic Press (AAP), Canada, Published Hard ISBN: 9781774639856.

25. Richardson AD, Hufkens K, Milliman T, Frolking S (2018) Intercomparison of phenological transition dates derived from the PhenoCam Dataset V1.0 and MODIS satellite remote sensing. *Scientific Reports* 8: 5679.

26. Liu S et al. (2017) Spatiotemporal dynamics of grassland aboveground biomass on the Qinghai-Tibet Plateau based on validated MODIS NDVI. *Scientific Reports* 7(1): 4182.

27. Zhang N, Wang M, Wang N (2002) Precision agriculture-a worldwide overview. *Computers and Electronics in Agriculture* 36(2–3): 113–132.

28. Sripada RP, Heiniger RW, White JG, Weisz R (2005) Aerial colour infrared photography for determining late-season nitrogen requirements in corn. *Agronomy Journal* 97(5): 1443–1451.

Index